LONG-TERM MONITORING

Why, What, Where, When & How?

Proceedings of a workshop and conference
"The Importance of the Long-term Monitoring of the Environment"
held by
Sherkin Island Marine Station
from 14th–19th September 2003
on Sherkin Island, Co Cork, Ireland

Edited by
John Solbé

MBE, DSc, CBiol, FIBiol, FIFM
St Asaph, Denbighshire, UK

Sherkin Island Marine Station, Sherkin Island, Co. Cork, Ireland
2005

This book is dedicated to the memory of

Professor G.E. (Tony) Fogg, FRS, CBE
1920-2005

Published by
Sherkin Island Marine Station,
Sherkin Island, Co. Cork, Ireland

Typeset and Layout by Susan Murphy Wickens

Printed by City Print Ltd., Victoria Cross, Cork, Ireland.

© Sherkin Island Marine Station 2005

All rights reserved. No part of this publication may be reproduced, stored in a retrieval system, or transmitted in any form or by any means, electronic, electrostatic, magnetic tape, mechanical, photocopying, recording or otherwise, without permission in writing from the publishers.

Cover photographs:
Whale fluke: International Whaling Commission
Sunset, Killary Harbour and gull: Robbie Murphy

Selected images and clipart: Copyright © 2005 Sherkin Island Marine Station and its licensors. All rights reserved.

ISBN: 1 870492 82 X

Printed on 50% recycled paper

PREFACE

SHERKIN Island Marine Station was founded in 1975 to study the ecology of our coastal waters by monitoring sites on the rocky shore of Sherkin Island and other shores in Roaringwater Bay. The initial data from the survey showed the need to expand our programme and today, it covers 700 miles of indented coastline Southwest from Cork Harbour to Bantry Bay. As we realised the importance of the long-term monitoring programme, we introduced companion surveys of phytoplankton, terrestrial plants, birds, insects and otters to expand our view of the ecosystem. The first of the Marine Station's long-term monitoring efforts is an example of what knowledge can do. The recently published work discusses 20 years of survey results of the rocky shores of Sherkin Island.

Matt Murphy, Director, Sherkin Island Marine Station

I have great difficulty understanding how people put so much faith in Environmental Impact Statements that are used for obtaining planning permission. At present, local authorities reviewing those studies do not have the tools to analyse and interpret environmental impact and make informed decisions. In most cases the data used to describe the natural conditions and project impacts are but snapshots of natural conditions that vary with the day and year. Often times the data are based on a few, brief site visits or less. These "views" of the environment lack full understanding and appreciation of natural variability and may even misrepresent the natural conditions, making them totally irrelevant.

The solution is better knowledge. I know that long-term monitoring provides that and is an absolute must. The proof of my belief is being revealed every day in our scramble to understand the earth's environmental shifts and how we respond. The mindset of those responsible for the protection of the environment must change if the environment is to be protected for future generations. For instance, a simple way to check on resource protections is to revisit projects and see if claims are supported by results. How many Environmental Impact Statements submitted to authorities are re-visited in later years? It is time to make it a condition of planning approval that projects are revisited and evaluated in a set number of years.

When I decided to host the Long Term Monitoring workshop and conference on Sherkin Island, I had many discussions with my friend and associate Michael Ludwig at the National Oceanic and Atmospheric Administration's National Marine Fisheries Services in the USA. He has similar views on long-term monitoring. With his wonderful support and advice we put together the agenda. In deciding on the criteria for participants in our workshop, we felt it was essential that the invitees had to have an understanding of, experience with and commitment to long-term monitoring.

Their challenge was to produce a reference and guide, a 'mini bible' for long-term monitoring. I believe that we have achieved that objective. No doubt some will say long-term monitoring is a luxury that is not financially affordable. I counter argue – our natural environment is a priceless heritage and must be protected. And, the price of protection is far less than the cost of loss.

The time and effort put into the workshop by the participant has been immense. Thank you to all of them for taking time out of their busy schedules to participate in it. They spent many long hours discussing and drafting their views, which have resulted in this book. They will, I know, forgive me if I make special mention of the eminent Prof. Tony Fogg for his great wisdom during the discussions.

I wish to thank my friend Prof. John Solbé for giving so much of his time in editing this book. He has skilfully brought together the workshop participant's views and experiences from a wide variety of careers and has presented it in manner that can be used by people in all walks of life. I am very grateful to John's wife Rosie, who gave much appreciated critical advice and assistance, particularly with the index of the book. I am continually grateful for the support of my family, without whom the work of the Marine Station could not survive.

Finally I must thank my daughter, Susan, who, together with John and Rosie, has put Trojan work into preparing the book for publication.

Matt Murphy, Director
Sherkin Island Marine Station

EXECUTIVE SUMMARY

What is long-term monitoring?

Long-Term Monitoring is the means by which we keep an eye on the environment to check whether defined management goals are being met. It requires observation, measurement, recording and analysis of organisms and the environment over suitable time-scales and in appropriate places. It allows us to look at our environment (either as a whole or at specific aspects such as a particular species) and decide whether we are maintaining it in a healthy state. It can provide early warnings of when things are going wrong and assist us in finding solutions when there are problems.

Management goals, precise or generic, set by managers working in conjunction with scientists and stakeholders can be of many types. Here are a few examples.

To ensure that:

- levels of particular chemicals do not exceed specific values in the tissue of indicator species, in air or water;
- the population size of species x is maintained at its present level;
- the populations of all species within the area are maintained in terms of number and distribution;
- the population size of species x recovers to its unexploited level.

Who benefits from long-term monitoring programmes?

The simple answer is that everyone (individuals, local communities, the global community, managers, industry, the environment itself) benefits from wise management decisions achieved through long-term monitoring programmes.

For example:

(1) Monitoring is an essential component of implementing the principles of sustainable development and ecosystem management, which are widely proposed at the global and local scales. Without long-term monitoring it is impossible to determine if an activity is indeed sustainable.

(2) It is vital to any meaningful review of Environmental Impact Assessments by providing a baseline of information to assess impacts, improving the quality of predictions of likely impacts and the ability to determine appropriate mitigation measures where necessary, such as in the siting of industrial plants, wind farms etc.

(3) Once a decision is taken, the predictions of likely impacts must be checked by monitoring programmes to identify if additional measures are required to meet stated management goals. An example might be where changes in fishing gear are implemented to reduce the bycatch of dolphins and porpoises.

(4) Industry will benefit from both the stability that arises out of sustainability and the better relationship with the wider community if it is shown that industry activities are not damaging the environment.

(5) People benefit from an improved quality of life, for example by an 'early warning system' of possible harmful changes to the environment and the identification of important factors in improving environmental quality. Improvements in air quality reducing the incidence of asthma and lung-related diseases would be one example.

How can we set up a long-term monitoring programme?

Designing a monitoring programme is not a simple task: it involves a number of steps before going into the field and implementing the programme. These are generally the same whatever the scale of the system being examined and might be considered as addressing the questions 'WHY? - WHAT?- WHERE?- WHEN? and HOW?'.

The steps are summarised below.

(1) Determine the management goals and try to quantify them.

(2) Decide what are the important factors to be monitored that will enable checks to be made on whether the management goals are being met. This will involve examining existing data and results from either the same areas/species or comparable areas/species and consulting relevant people or groups in the area.

(3) Consider methods of collecting relevant data and determining the probable quality of those data given the availability of resources (in terms of practicality, money and expertise). This may involve carrying out a feasibility study.

(4) Given the quality of data believed obtainable, consider the means of analysis of the data for changes or trends. Assume various levels of sampling (both in terms of distribution and frequency), and determine the degree of confidence in the ability to detect changes, should they occur. This will usually involve examining statistical confidence, given various assumed trends.

(5) From the above, determine the minimum sampling requirements (e.g. in terms of important factors to measure, sampling design in time and space) to be able to address the objectives. Remember that defining a minimum does not mean that aiming higher is prohibited - but it will prevent wasted effort.

Are long-term monitoring protocols flexible?

The essential ingredient of a long-term dataset is that the data are comparable across time, so that if a change is detected, there can be confidence that the monitoring methods are not responsible for the change. Thus if improvements (e.g. in terms of equipment, methods including sampling design, analytical techniques or personnel) are introduced, sufficient care must be taken to ensure that calibration of any changes is undertaken. One way of achieving this is to run old and new methods in parallel for a period of time.

Who can collect data?

The important principle is that data quality is maintained, not who collects the data. Where practicable, there is great value in involving 'stakeholders' (such as local communities, fishermen or managers) in the monitoring process. With appropriate training they could help to collect data or to provide regular explanations of the programme and its results.

Is long-term monitoring predictive?

Once management goals have been established, it may be important to identify the cause or causes of any changes observed, or to predict the effect of certain external changes such as what might happen if a factory was built nearby. Initially, a long-term monitoring programme can provide a baseline for evaluating whether change *has* occurred. As the programme continues and data accumulate, they can be used to establish plausible explanations of changes. Further, they can give sufficient insight to allow *predictions* of the likely effects of changes. Perhaps most importantly, monitoring allows the *value* of the predictions to be determined and enables the development of even better predictive capabilities.

Is long-term monitoring an optional extra?

NO! It is essential and pressure must be put on the relevant authorities to ensure that well-designed programmes are either maintained or put into place. There is also a responsibility on scientists to make available data that may be valuable in long-term monitoring programmes, even if the data were not collected for that purpose in the first place.

The report of the Workshop and the papers given at the Conference now provide the detail needed for further understanding of the methods, uses for and examples of long-term monitoring.

(Based on text provided by Greg Donovan. Ed.)

Contents

PREFACE iii

EXECUTIVE SUMMARY v

1. **INTRODUCTION** 1

2. **OUTPUT OF THE WORKSHOP** 3

    Editor's Note on the Workshop Report 3

    Participants in the Workshop 4

    Workshop Report 5

2.1 **What is Long-Term Monitoring?** 5

    a) Definitions: surveillance and monitoring; long-term and short-term monitoring 5

    b) The reasons for and benefits of establishing a programme of long-term monitoring .. 9

    c) Opportunities for monitoring by amateur individuals and groups 11

2.2 **Creating a Long-Term Monitoring Programme** 13

    a) Aims and objectives 14

    b) The material to be observed 15

    c) Location and collections 15

        i) Sampling sites 15
        ii) Biological samples 16
        iii) Reference sites and reference collections 17

    d) Sampling frequency and duration 19

    e) Sampling methods 20

    f) Trial periods; sampling limitations; damage from sampling 20

2.3 **The Data** 22

    a) Sample design: statistical treatment 23

	b) Accuracy and precision; quality control 23
	c) Relationships between sampling frequency and intended applications 24
	d) Who can collect data? 25
	e) Quality control 25
	f) Accessibility and other issues 26
	g) Using data sets not originating in monitoring programmes 26
2.4	**Using the Results** **28**
	a) Interpretation – general 28
	b) Monitoring compliance standards 29
	c) Prediction 30
	i) Prediction within the sampled area 30
	ii) Prediction outside the sampled area 30
	d) Examples (and benefits) of long-term monitoring 30
	e) Communication 31
2.5	**Recommendations from the Workshop** **32**
	a) New programmes 32
	b) Existing or historic programmes 32
	c) Special Recommendations 32
3.	**PAPERS FROM CONFERENCE** **34**
	Editor's Introduction to the Conference Papers 35
3.1	**Opening Address** **36**
	Pat the Cope Gallagher TD, Minister of State at the Department of the Environment, Heritage and Local Government, Dublin, Ireland
3.2	**Is Short-Term Monitoring Sufficient?** **40**
	Prof G.E. (Tony) Fogg, Prof Emeritus in Marine Biology, School of Ocean Sciences, University of Wales, Bangor, UK
3.3	**An Overview of the Social Value of Long-Term Monitoring** **44**
	Prof John F. Solbé, Environmental Consultant, St Asaph, Wales, UK

3.4 Long-Term Monitoring – A Media Viewpoint 51

Alex Kirby, Environmental Journalist, UK

3.5 The Role of Geology in Long-Term Monitoring 56

Dr Peadar McArdle, Director, Geological Survey of Ireland, Dublin, Ireland

3.6 Thirty Years Monitoring Waters, Weeds and Fishes 66

W.S.T. Champ; M.F. O'Grady; P. Gargan; P. Fitzmaurice and P. Green, Central Fisheries Board, Dublin, Ireland

3.7 Long-Term Monitoring of Birds in Ireland 79

Oscar J. Merne, National Parks & Wildlife Service, Department of the Environment, Heritage & Local Government, Dublin, Ireland

3.8 Otters and Fish Farming – A Good News Story 87

Jane Twelves, Salar, Isle of South Uist, Outer Hebrides, Scotland, UK

3.9 Thirteen Years of Monitoring Sea Lice in Farmed Salmonids 92

Dr David Jackson, Lorraine Copley, Frank Kane, Oisín Naughton, Suzanne Kennedy and Pauline O'Donohoe, Marine Institute, Galway, Ireland

3.10 The Sedimentary Record Shows the Need for Long-Term Monitoring of Phytoplankton 106

Dr Barrie Dale, Department of GeoSciences, University of Oslo, Norway

3.11 Rocky-Shore Monitoring at Sherkin Island Marine Station since 1975 114

Dr Gillian Bishop, Environmental Consultant, Aberdeen, Scotland, UK

3.12 Long-Term Fisheries Monitoring with Emphasis on the Striped Bass (*Morone saxatilis*) from the Hudson River 132

Byron Young, Kim A. McKown and Julia M. Brischler, New York State Department of Environmental Conservation, USA

3.13 Long-Term Monitoring of Marine Phytoplankton at Sherkin Island Marine Station 140

Dr Geraldine Reid, Curator of Diatoms, Natural History Museum, London, UK

3.14 Long-term Observations: Crustaceans and Molluscs in Atlantic Canada 149

Dr René Lavoie, Fisheries & Oceans Canada, Nova Scotia, Canada

3.15 Shellfish Toxicity in the NW Atlantic: Unexpected and Widespread Occurrences of *Alexandrium fundyense* Balech in Coastal New England, September 1972. Where was the monitoring? 157

Christopher Martin, NOAA/National Marine Fisheries Service, Milford, CT, USA

3.16 Cetaceans: Can we Manage to Conserve them? The Role of Long-Term Monitoring.. 161

Greg Donovan, International Whaling Commission, Cambridge, England, UK

3.17 Environmental Monitoring in Ireland: Aspects of the Role of the Environmental Protection Agency 175

Larry Stapleton, Environmental Protection Agency, Wexford, Ireland.

3.18 The Importance of Long-Term Monitoring of the Environment 185

Michael Ludwig, National Marine Fisheries Service, Milford, CT, USA

REFERENCES 202

INDEX 215

1.
INTRODUCTION

People are concerned by frequent reports of environmental deterioration with possibly life-threatening consequences for human societies:

- **Global warming** – threatens to cause increasing droughts in some parts of the world, increasing storms and flooding in others, and ultimately ice melting at the poles raising sea levels, endangering millions of people in low-lying coastal regions.
- **Ozone depletion** – threatens to increase ultraviolet radiation leading to increased cancer development in humans and a variety of detrimental effects on other forms of life.
- **Pollution** – The increasing human population inevitably increases all forms of pollution: human waste causes eutrophication and contamination of freshwater and coastal environments, while industrial waste and energy requirements cause chemical pollution of air, ground and water; also shipping causes noise pollution in the oceans.
- **Destruction of natural habitats** – Human encroachment is depleting natural stocks of plants and animals, leading to reduced numbers of species locally, and upon the Earth.

The consequences to humans of this environmental deterioration are enormous. In addition to the more obvious life-threatening consequences, we also face severe economic and societal consequences. On the economic front, we face increasing expense as it becomes more and more costly just to maintain safe standards of drinking water and food production, while at the same time there are huge losses due to the costs of cleaning up pollution. Even societies wealthy enough to pay at least the local costs of environmental deterioration are nevertheless affected in other ways, and deteriorating quality of life is increasingly a main driving force for the richer nations to invest more effort into environmental protection.

One major factor fuelling public concern over environmental deterioration is the uncertainty surrounding the main issues. Ideally, we would tackle these issues of changing environments by first understanding the natural environment sufficiently in order to identify the change. In truth we are facing the urgent need to understand changes to natural systems when we have little understanding of their former state. There are several main reasons for our being inadequately prepared:

- **Natural systems are complex** – understanding them is therefore difficult, demanding enormous scientific efforts on a larger scale than has previously proved possible.

- **Natural systems are characterised by a high degree of variation** – on time-scales of hours to thousands of years; understanding this requires observations spanning at least the time frame of interest to humans (up to several hundreds of years).
- **We lack the necessary long-term observations** – to adequately understand the background for assessing the perceived threat from environmental change.

For these reasons, we are forced into making estimates of environmental change in the absence of adequate background information. In practice this involves using available, mostly short-term, observations to answer questions requiring longer-term series of data, with a heavy reliance on modelling to help make up for the shortfall. This inevitably produces large amounts of uncertainty in any predictions generated by the models, and it is this that adds to the uncertainty felt also by the public.

The issue of global warming illustrates this. The curves showing the predictions of global warming used by the International Climate Panel and others making decisions usually include measurements from about 1960 to the present, to which are added a mathematically generated predicted curve for future global warming. The level of uncertainty is usually drawn in as an upper curve representing the most extremely high degree of warming considered possible by the model, and an equivalent lower limit defined statistically. The mid-curve defining the mean suggests predicted global warming is often used as a starting point for policy making decisions, but the wide spread of possibilities defined by the upper and lower limits of uncertainty allow for very different scenarios, with potentially alarming results for future generations of people. Our best hope is that the truth lies closer to the lower limits of variability which are closer to a reasonable increase of temperature, closer to that experienced in the historical past, for example during Mediaeval times.

All this raises the question of what can we do to reduce the levels of uncertainty regarding environmental issues? The obvious need for more longer-term observations cannot be met instantaneously. However, if more such records would have helped us to understand present-day environmental changes and improved our ability to predict future changes, as seems to be the case, we should at least consider the possibilities for maintaining the few long-term series of observations we have in place and starting new series where needed. This was the subject of an international workshop leading to the publication of this book.

Barrie Dale

2.
OUTPUT OF THE WORKSHOP

EDITOR'S NOTE ON THE WORKSHOP REPORT

Barrie Dale's INTRODUCTION has set the scene for this account of a Workshop and Conference on the needs for and value of long-term monitoring in our responsibility to care for our environment locally, regionally and globally. The event was conceived and organised by Matt Murphy, Director, Sherkin Island Marine Station and brought together experts from academia, government and industry with experience in various branches of science and the media. What they had in common was a passionate interest in long-term monitoring and the benefits it can bring to a soundly based debate on environmental protection and sustainable development.

The Workshop was held in 'The Islander's Rest' on Sherkin Island and we are grateful for the excellent care provided by Mark Murphy and his staff. The beautiful setting overlooking Baltimore Harbour only served to emphasise what could be at risk if we do not take proper care of our environment.

The proceedings were in two distinct parts: an intensive three-day Workshop attended by a small number of experts and a two-day Conference attended by a much broader range of stakeholders.

Section 2 of this book resulted from the concluding session at the Workshop in which all the participants contributed in discussions and by providing written contributions. The proceedings of the Workshop were admirably steered by Michael Ludwig and Gillian Bishop. The Editor's task has been to assemble this valuable but diverse material into the structured summary below, taking into account (as several contributors have mentioned) that in order for a publication to be a success it must strike the right note between scientific detail and a form of language which will attract and hold the attention of busy people.

Among other contributions in Section 2 may be found anecdotes and small historical examples, which serve to demonstrate different aspects of long-term monitoring. Several important messages may also be found, which, if not acted on (for example by academic institutions) will deprive those responsible for the sustainable use of natural resources of essential techniques and expertise.

Thus it is hoped that by reading Section 2 the interested reader may be able quickly to assimilate a summary of the nature and benefits of long-term monitoring for the environment. Those needing more detail will be able to pursue their enquiry using Section 3 of this book, particularly employing the Index, which has been deliberately drawn up to be more extensive than normal.

John Solbé

Participants in the Workshop

Name	Affiliation
Dr. Gillian Bishop	Environmental Consultant, Aberdeen, Scotland.
Mr. Trevor Champ	Central Fisheries Board, Dublin, Ireland
Prof. Barrie Dale	Dept. of Geosciences, University of Oslo, Norway.
Mr. Greg Donovan	International Whaling Commission, Cambridge, UK.
Prof. Tony Fogg	Professor Emeritus in Marine Biology, University of Wales, Bangor, UK.
Dr. David Jackson	Marine Institute, Galway, Ireland.
Mr. Alex Kirby	Environmental Journalist, UK.
Mr. Padraic Larkin	Director, Environmental Protection Agency, Wexford, Ireland.
Dr. René Lavoie	Fisheries & Oceans Canada, Nova Scotia, Canada.
Mr. Michael Ludwig	National Marine Fisheries Service, Milford, CT, USA.
Mr. Chris Martin	National Marine Fisheries Service, Milford, CT, USA.
Dr. Peadar McArdle	Geological Survey of Ireland, Dublin, Ireland.
Mr. Oscar Merne	National Parks & Wildlife, Dept. of the Environment, Heritage and Local Government, Dublin, Ireland.
Mr. Robbie Murphy	Sherkin Island Marine Station, Cork, Ireland.
Ms. Susan Murphy Wickens	Sherkin Island Marine Station, Cork, Ireland.
Dr. Geraldine Reid	Curator of Diatoms, The Natural History Museum, London, UK
Prof. John Solbé	JF Solbé – Environmental, Denbighshire, UK.
Dr. Rosie Solbé	Institute of Biology, North Wales, UK.
Mr. Larry Stapleton	Environmental Protection Agency, Wexford, Ireland.
Mr. Eric Twelves	Salar, Isle of South Uist, Outer Hebrides, Scotland.
Ms. Jane Twelves	Salar, Isle of South Uist, Outer Hebrides, Scotland.
Mr. Koen Verbruggen	Geological Survey of Ireland, Dublin, Ireland.
Mr. Byron Young	NY State Dept. of Environmental Conservation, New York, USA.

Participants in the Workshop: *from left*, John Solbé, Rosie Solbé, Susan Murphy Wickens, Alex Kirby, Jane Twelves, Greg Donovan, Dave Jackson, Trevor Champ, Barrie Dale, Eric Twelves, Tony Fogg, René Lavoie, Geraldine Reid, Michael Ludwig, Larry Stapleton, Alloys Martin, Chris Martin, Koen Verbruggen, Gillian Bishop, Byron Young, Matt Murphy, Dale Young. Inset: Robbie Murphy (photographer)

WORKSHOP REPORT

2.1

WHAT IS LONG-TERM MONITORING?

a) Definitions: surveillance and monitoring; long-term and short-term monitoring

Definitions: Strictly speaking, monitoring and surveillance are not terms we can use interchangeably. Monitoring is just one aspect of the broad area of activity described by the word surveillance. Surveillance means "keeping a watch over". A formal definition of surveillance for environmental objectives is that it constitutes "a continued programme of surveys systematically undertaken to provide a series of observations in time" (Hellawell, 1978). It doesn't say that you have to have some pre-conceived idea of the result of your observations. It may be that you observe and record things out of curiosity and a feeling that they could be interesting or useful. Subsequently you may see what use can be made of them. Alternatively, of course, you may have a very clear idea of an issue, and a programme which you think may be valuable in addressing it. Examples follow.

Land Lying Fallow

Virgil, writing 'The Georgics' before 19 BC, advised farmers to let land lie fallow in alternate years. This sensible advice will have originated in careful observation of various possible practices, where the outcome, as far as we know, was not predicted by any hypothesis before the 'experiment'.

A Naturalist's Diary

In the eighteenth century the Revd Gilbert White of Selbourne wrote about many of the natural phenomena he observed in his parish, without any particular agenda. By keeping his diary he could note the differences from year to year in the weather, the arrival and departure of various species of bird, first flowering of spring plants and so on. He probably did not do this because he thought there were progressive changes from year to year, but simply to record and compare – and postulate. He was undertaking surveillance.

> **Marine Plankton Collection**
>
> Alister Hardy in the 1920s noticed variations in the catches of marine plankton. He realised that a lot of information could be gathered about the distribution and abundance of plankton by using merchant ships on regular trade runs, so with this aim he designed the Continuous Plankton Recorder (Hardy, 1956). The knowledge gained by building this huge data set on plankton over the decades is being used to help diagnose the causes of present-day events in terms of such issues as climate change, pollution and exploitation.

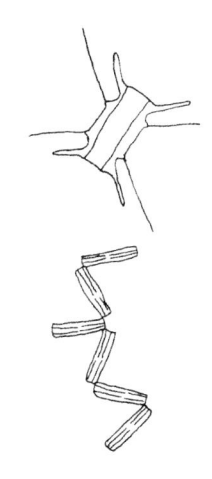

What distinguishes *monitoring* from this general definition is that monitoring is *surveillance undertaken to ensure that previously formulated standards are being met.*

For example, there are regular monitoring programmes of bacterial contamination of bathing waters, chemical contamination of drinking water and the air, and so on. Generally the observations are set against an agreed standard, set by an authority such as the World Health Organisation, the European Commission or a national authority. The result of the monitoring programme will be a report of the measurements made, with an indication of whether the samples have passed or failed. Typically too there may be a statement of the trends which the data may be demonstrating, either of improvement, deterioration or a static situation. All of this is covered by the strict definition of monitoring.

In recent years, however, the more precise definition of monitoring has been lost. Therefore, in this book, the term monitoring can be taken to refer to both its restricted meaning and to general surveillance. Where a distinction is needed it will be made afresh. Thus, in long-term monitoring we can envisage a single monitoring programme or maybe a series of programmes which give rise to sets of environmental information in some kind of a similar format and which span (for the present purpose) a substantial period of time.

Duration: (See also 2.2 (d).) Of course, a monitoring campaign can be defined as short-, medium- or long-term, and this distinction is rather arbitrary. We could for example monitor the many generations of aphids (greenfly) throughout a summer, or the same number of generations of whales over many decades. In both cases what we are doing can be seen as the continuous sampling (at appropriate temporal and geographical levels and scales) of organisms and/or their environment in response to defined *management goals*, which should be clearly stated. In the first case it might be to do with the protection of the crop of a single season so that remedial action could be taken should aphids exceed levels at which crop yields would be reduced; in the second it might be part of many years work to establish the status of an endangered species. Trends in abundance would be measured to examine whether mitigation measures established to ensure that the population began to increase were working. In both cases we do not necessarily have a fixed standard in mind, so these are examples of surveillance rather than monitoring, but as stated above we are not concerning ourselves with the niceties of terminology here.

This book is concerned with demonstrating the essential and integral role well-specified and designed monitoring programmes play in our attempts to manage human activities such that they are sustainable, and allow for a healthy and varied local and global environment with the full range of biodiversity.

Geographic Scale: An appropriate geographic scale as well as a temporal scale ('duration') must be chosen in a monitoring programme. The following examples demonstrate a combination of geographic and temporal scales, and there are many others throughout this book.

Example 1: Global – Climate monitoring

>The climates of the Earth change, and always have, from day to day, with latitude, longitude, season and from year to year. At present we are concerned with global warming and the way this correlates with increasing levels of (mostly) carbon dioxide in the atmosphere, creating a 'greenhouse' effect. There have been many warmer periods in the Earth's history and many colder periods, but most scientists now accept that the current warming is closely connected with our combustion of large amounts of fossil fuel at rates faster than plants can utilise the resulting carbon dioxide from the air and, by photosynthesis, lock away the carbon in their tissues, and away from the atmosphere, for some period of time. There are global networks of research laboratories and weather stations monitoring the possible causes and the effects of these climatic events.

Example 2: Regional – Water Framework Directive

>The European Union has recently put in place the 'Water Framework Directive' (European Council, 2000). The role of this new piece of legislation, which gathers up many preceding laws and regulations for its purpose, is to work towards excellent water quality (fresh waters, estuaries and coastal waters) throughout the EU. In order to achieve this a system of monitoring will be established to define the current state of aquatic ecosystems, seek the causes of any quality less than 'good' and put in place management systems to bring the situation up to 'good', (or even better 'excellent'). Implementing this will necessitate setting up long-term monitoring sites in representative locations and sampling at regular intervals according to standardised guidelines. This is truly monitoring rather than surveillance, because the EU has already defined in general terms what it means

by 'excellent', 'good', 'moderate' and 'poor' quality in ecosystems, separating out descriptions of, for example, the fish community, invertebrates and plants.

Example 3: Local – Garden bird census

Many groups of birdwatchers, amateur ornithologists, have traditionally carried out long-term monitoring programmes (eg BirdWatch Ireland, and in the UK, under the *aegis* of the British Trust for Ornithology or the Royal Society for the Protection of Birds) by recording the numbers and species of birds in a given area, perhaps over a certain weekend or over a full year.

b) The reasons for and benefits of establishing a programme of long-term monitoring

"Using present data to connect past reality with future reality"

There are very few, perhaps no, aspects of the Earth which are static. Some changes occur very slowly. (The movement of tectonic plates 'Continental Drift' is said to occur at about the same pace as the growth of our toenails, at least for the rate of widening of the Atlantic.) But many changes occur extremely rapidly (weather systems) or suddenly (earthquakes) and do not tend towards any state of true, long-term stability. Because our world is not static, and we want to be sure that we understand the causes of changes, we need to acquire facts about the environment now, as well as trying to interpret evidence for past events. This requirement is not idle curiosity: it can help us to foresee environmental and economic disasters and give us clues on how best to prevent them, delay them or lessen their impact.

Everyone (individuals, local communities, the global community, managers, industry, the environment itself) benefits from wise management decisions and long term monitoring programmes are central to this. They are essential to local, national or international management of human activities, from predicting impacts at the planning stages to enable wise decisions to be taken, to ensuring the adequacy of the predictions and the effectiveness of mitigation measures to protect the environment, once decisions have been taken. More specific examples of the need for and benefits of such programmes are given below.

1. Monitoring is an essential component of implementing the principles of *sustainable development and ecosystem management*, which are widely proposed at the global and local scale. Without long-term monitoring it is impossible to determine if an activity is indeed sustainable.

2. Monitoring is vital to any meaningful review of *environmental impact assessments and statements* (EIA/EIS), in that it provides a baseline of information to assess impacts, improves the quality of predictions of likely impacts and the ability to determine appropriate mitigation measures, where necessary. An EIA can be simply a compliance exercise. Properly done, it can achieve worthwhile results as a one-off snapshot of a short-term reality. But it can also serve to launch a longer-term

monitoring exercise which may turn into fully fledged long-term monitoring. If it is well done, then at best it can provide a prediction of the likely impact of the project based on the existing data of the 'current state of the environment' and the associated natural variation. It therefore must lead to a monitoring programme to assess whether the predictions, however well founded, are correct.

3. *Establishing the consequences of management decisions* – It is essential that when we make a management decision based on predictions of likely impacts on the environment (with or without the benefit of existing monitoring data) we follow up the result of the decision with a suitable monitoring programme. From this process we can check the accuracy of our predictions and identify any areas for improvement when we next have to make such decisions.

4. *Demonstrating sustainable development* – Where an industrial company, a local authority or any other body has introduced a programme or process with the stated aim of improving the sustainability of their operations it will benefit the originator if it can be shown by monitoring that they are succeeding in their aim. Society and the originator will gain from the consequent stability (or improvement) that arises out of an intended sustainable development. Better relationships with the wider community must result if it can be shown by monitoring that activities are not only *not* damaging the environment but are actually improving the situation.

5. *Early warning systems* – Long-term monitoring is one of the ways in which early warning may be obtained of an impending environmental problem, for example a volcanic eruption, changes in sea-level etc.

6. *Post-hoc awareness of problems* – The classic modern example is related to the use of persistent and bioaccumulating pesticides, where the loss of song birds alerted society to this major problem of previous decades (Carson, 1962).

7. Long-term monitoring can also provide "*a standard for mitigation*" – a baseline for the recovery of a damaged habitat, by showing us how far we have to go to achieve the restoration of what has been lost or creating habitats to replace what cannot be restored. This concept of "no nett loss" has been applied in Canada. In the EU there is a policy of no nett loss of wetlands.

The benefits of long-term monitoring are apparent from the examples above, but we can try to place these in a logical order (see Table 2.1).

Let us end this section with another mantra of long-term monitoring.

"The best way of planning for the future is by understanding the past and how it has shaped our present."

Table 2.1: Benefits of long-term monitoring

Sequence	Value of long-term monitoring
Site-specific, eg an industrial development or new waste treatment regime	
At the planning stage ie predictive	Demonstrate original condition of site/area. (Even the planning stage can take some years, so there is a real chance of obtaining more than one year's data.)
During construction / implementation	Follow consequences in real time and take remedial action where necessary. Monitor for compliance with a standard.
During use/production phase	Follow consequences in real time and take remedial action where necessary. Monitor for compliance with a standard.
After site closure or end of project	Record any recovery and any other consequence of closure. Monitor for compliance with a standard.
Widespread and long-lasting, eg oceanic or atmospheric changes	
As part of a programme of surveillance with no preconception of issue	Observations may result in hypotheses concerning an environmental change or other issue.
Surveillance targeted at an issue	Observations may test hypotheses and refine them, eventually leading to a deeper understanding of the issue, and to remedial measures.

We can compare past realities that have similarities with the present to inform the decisions we make. If we have insight, we are likely to make decisions built on surer foundations.

c) Opportunities for monitoring by amateur individuals and groups

The idea of long-term monitoring can seem overwhelming when terms and information are presented. However, when the actual sample items and procedures are fully described, it is not unusual that the work is nothing special and the people involved, although trained and committed, need possess no unusual skills.

Anyone can start to record observations of the environment and collect data which may develop into a long-term monitoring programme. People have been collecting data sets relevant to environmental monitoring for centuries without realising it. We now recognise that these have provided us with a valuable record of environmental change. (For example vineyard records going back to the Middle Ages are used as evidence of long-term climatic change.) Nowadays monitoring programmes can be scientifically designed to give clear indicators of environmental status and change.

There are any number of different schemes that might collect information that can be used in reporting details about an organism and the environment. Because of the wide range of possible subjects, a long-term monitoring project can be created by anyone, provided they follow the basic guidance set out in this discussion (See 2.2 – 2.4). Useful records of the dates of flowering, the arrival of migratory birds or the loss and recovery of peregrine falcons have been used to call attention to events or changes. In some cases the records began with simple observations that gained in importance as they grew in longevity.

Because of the importance of weather changes, many individuals from all walks of life have carried out local and regional long-term monitoring projects. While these records have been important in understanding climatic change, we have missed opportunities for more complete understanding because of differences in how we collect and consider the evidence. This is changing as both trained and untrained record keepers have been brought together or exposed to sampling ideas, methods and the centralised databanks.

Examples of 'amateur' work at the local level

1. Ornithology: Many groups of so-called amateurs have traditionally carried out long-term monitoring programmes (eg BirdWatch Ireland, British Trust for Ornithology / Royal Society for the Protection of Birds censuses of birds, etc).

2. Phenology: Phenology is the name given to the study of the dates in the year when natural phenomena recur from year to year. For example it could be noting when you heard the first cuckoo in Spring (a phenomenon giving rise to letters to the newspaper over the decades)! Or it could be the date when you saw natural snowdrops first flower, or noted your first and last sightings of swallows or house martins in the year. Nowadays, people interested in phenology are commonly linked through websites such as http://www.phenology.org.uk/newsletter.htm. Their own data are added to the national data so that, for instance, you can see how Spring is developing from south to north. In relation to climate change, the data examined have to be part of a long series. The British phenology site, as well as describing the start of phenology in Britain by Robert Marsham with his 'Indications of Spring' as early as 1736, also refers to ancient records in Japan and China concerned with the time of blossoming of cherry and peach trees back to the eighth century. In 1875, the UK website states, British phenology took a major leap forward when the Royal Meteorological Society established a national recorder network. Annual reports were published up until 1948.

Such efforts are increasingly being recognised as a valuable contribution to environmental monitoring. These days there are people in society who are interested in taking part as amateurs and volunteers in environmental monitoring programmes and this should be encouraged enthusiastically. However, it is important that the principles of good design, quality control and consistency are met. This places responsibility on amateurs wishing to carry out such a programme to ask for advice and for scientists to share their expertise when required and to consider the use of dedicated local volunteers when designing monitoring programmes.

2.2
CREATING A LONG-TERM MONITORING PROGRAMME

Long-term monitoring can be seen as the continuous sampling (at appropriate temporal and geographical levels and scales) of organisms and/or their environment in response to defined management goals. These management goals (which may be precise or generic) should be set by managers in conjunction with scientists. In simple terms, monitoring allows us to look at our environment (either as a whole or specific aspects such as a particular species) and decide whether we are maintaining it in a healthy state. It can provide early warnings when things are going wrong and assist us in finding solutions when there are problems.

Designing a monitoring programme is not a simple task and involves a number of steps before going into the field and implementing the programme (Figure 2.1). These steps are generally the same whatever the scale of the system being examined and might be considered as answering the questions 'WHY? - WHAT?- WHERE?- WHEN?- HOW?'. These steps are summarised below.

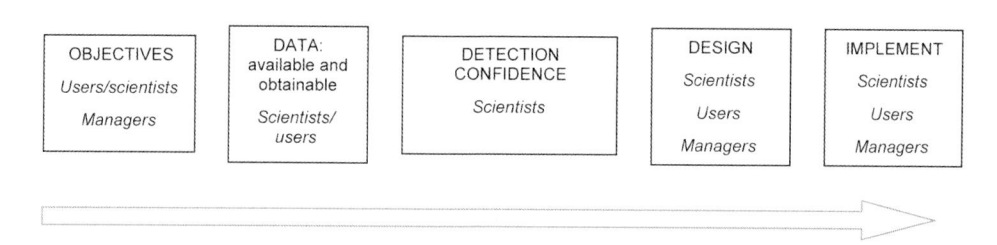

OBJECTIVES	DATA: available and obtainable	DETECTION CONFIDENCE	DESIGN	IMPLEMENT
Users/scientists		Scientists	Scientists	Scientists
Managers	Scientists/ users		Users	Users
			Managers	Managers

Figure 2.2: The steps to a monitoring programme

(1) Determine the management goals and try to quantify them.

(2) Decide what are the important factors to be measured/monitored that will enable you to find out whether the management goals are being met - this will involve examining existing data and results from either the same areas/species or comparable areas/species and consulting relevant people or groups in the area.

(3) Consider methods of collecting relevant data and determining the probable quality of those data given your likely resources (in terms of practicality, money and expertise) - this may involve carrying out a feasibility study.

(4) Given the quality of data you believe you can obtain, consider how you might analyse those data for changes or trends. Assume various levels of

sampling (both in terms of distribution and frequency), and determine the confidence you might have that you would detect changes should they occur - this will usually involve examining statistical confidence given various assumed trends.

(5) From the above, determine the minimum sampling requirements (e.g. in terms of important factors to measure, sampling design in time and space) to be able to address the objectives. Remember that defining a minimum does not mean that you should not aim higher - but it will stop you putting in wasted effort.

The following sections discuss these steps in more detail.

a) Aims and objectives

At the outset it is essential to be totally clear on the reasons for setting up a monitoring programme. The importance of this stage cannot be over-emphasised. It is impossible to design a good monitoring programme (or indeed any scientific programme) without a clear statement of the management goals it is supposed to be monitoring. These management goals (which may be precise or generic) should be set by managers (e.g. local authorities, national governments, international organisations) in conjunction with stakeholders and scientists. The consultation with stakeholders (e.g. local communities, businesses, fishermen, NGOs, research groups etc.) is essential if an atmosphere of trust is to prevail. It may well be worth considering setting up a Steering Group including representatives of the various stakeholder groups to oversee the whole process.

It is best to co-ordinate the programme design with as many knowledgeable individuals as possible. Because the question(s) being asked are often from management, there should be members from that group as well as informed ecologists and statisticians. Their challenge is to identify and account for the factors that might vary, such as growth and reproduction and to make allowances for these in relation to whatever the specific objectives of the long-term monitoring might be. It is recommended that a trial period be used to ensure that errors and mistakes are minimised.

In the sequence: "WHY? – WHAT? – WHERE? – WHEN? – HOW?" the *aims and objectives* must address and record answers to the first question: "Why?"

Although there will be many underlying principles in common, there is no single answer which can be written down to cover aims and objectives for all cases, as can be observed by looking at the wide diversity of monitoring examples described in this book or in the general notes in 2.1(b) above. Each programme's objectives will require careful and individual attention. All 'stakeholders' with an interest in the programme (eg local communities, businesses, fishermen, NGOs, research groups, intergovernmental organisations, national or local governments) should be consulted. They could consider setting up a Steering Group to be involved for the duration of the programme so that no group feels left out. At the very least, explanation to the stakeholders of what is happening at each stage of the process, should be undertaken. The goals themselves will be case-specific and may be expressed in terms of individual species, groups of species, specific aspects of the environment or the environment as a whole. The importance of any management goal is that it should be quantifiable in some way so that an appropriate monitoring programme can be designed. *If the "Why?" is not clear at the outset to everyone concerned with establishing the programme, following it through, interpreting the results and acting on those results, a great deal of time may be wasted further down the line in modifying the programme to address satisfactorily the original question posed.*

b) The material to be observed

Here, we are considering the question "What should be measured?" The answer to this will depend on the reasons for monitoring, the 'management goal', as explained above. Clearly there will be as many answers as there are problems, but we need to establish a few principles. It will not only be new information which can be considered: there can be great value in historical data, in the views of local people such as the observations of fishermen and in the results of monitoring programmes for other purposes in the area.

Measurements must be meaningful and, wherever possible, provide unequivocal data. The word 'meaningful' implies concepts such as 'reliability' and 'relevance'. If samples, for example of seawater or of seal fat, are taken for chemical analysis, the analytical methods used must be capable of detecting relevant substances at appropriate concentrations with known degrees of precision (inter-measurement variation) and accuracy (proximity to the real value). If samples are of plankton, the larger algae, invertebrates or fish, the identification of the appropriate taxonomic level must be possible and any relevance of the different behaviour of life-cycle stages must be clearly understood. For both biological material and other samples it may be valuable to have standards available to check the accuracy of findings. This is considered further in the following section.

c) Location and collections

Under this heading we not only need to address the question "Where should measurements be made?" but also consider the value of reference sites, so that results at sampling sites can be placed in context.

i) Sampling sites

Where the reason for monitoring is connected with some new development or other anticipated change of an area it is obvious that some sampling sites should be located in that area, others nearby, and some further away. The locations should be chosen to take into account where we would be most able to allow the data collected to be used to determine whether the management goals are being met. For example, if it is expected that a point-source discharge into a river needs to be monitored, sampling locations a) upstream, b) at the discharge point and c) downstream may be needed, taking into account such matters as river flow, mixing zones and time of travel. If the interest is in some local change in air quality, the prevailing direction and likely strength of the wind can be used to set sample sites (and the heights above ground level where air is to be sampled). A coastal problem may best be sampled with regard to tidal and long-shore flows.

Establishing a so-called 'transect' – a line of sampling positions in land, air or water – can be a valuable way of demonstrating trends, whether natural (rocky shore ecology changing with height above the low-water mark) or man-induced (distance downstream of a smelter).

For large-scale issues such as the position of the Gulf Stream, global climate changes or the loss of tropical rain forest, the same principles may apply, but on a massive spatial scale and often over a long period of time (see 2.2 (d): Sampling frequency and duration).

ii) Biological samples

At the heart of many long-term monitoring programmes is a collection of biological material. This material may be collected, examined, data derived and the sample then discarded. However, it is frequently vital to retain samples for future purposes. To ensure that the work moves forward and continues to provide useful information it is essential that the labelling and archiving of samples be considered initially and practised faithfully. An example of failure to perform this task can be found in any number of locations where identification tags or records are lost, misplaced or illegible, making the sample virtually worthless. The techniques used to label samples have improved to avoid these problems. They should be used. Another vital tool for programmes is the specimen collection or *reference collection* (see below). Having identified specimens to refer to when uncertainties arise has repeatedly eliminated problems. Identification books and the availability of experts in identification are other components of the support structure for sampling programmes.

Great attention must be paid to storage conditions, so that both the identity and the

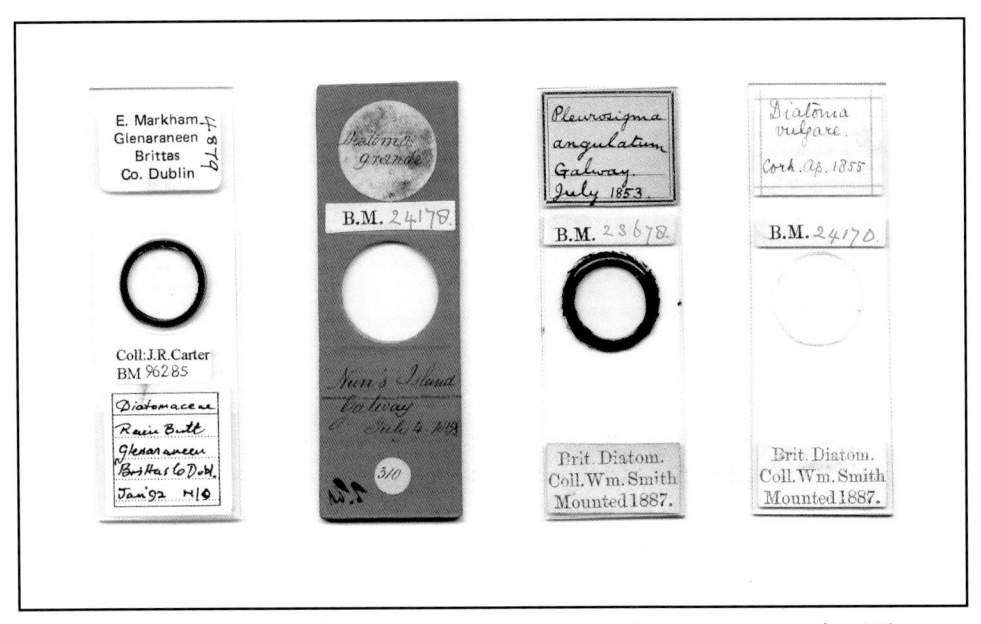

Clearly labelled specimens (photograph courtesey of the Natural History Museum, London, UK)

'integrity' of the sample are retained. 'Integrity' here means that there is no change of or damage to the sample. Storage conditions to achieve this constant state of the sample may be defined in terms of temperature, moisture content, lighting, preserving fluids etc.

Budgeting for specimen and sample collection and maintenance in storage should be included in planning at the outset of the monitoring programme.

As well as their value for the monitoring exercise, collections of samples and specimens may have a particular role to play in maintaining continuity in a long-term programme, even though scientific staff may change, new analytical methods may be developed and taxonomic nomenclature, or even classification, may evolve. (An example of a change in name was demonstrated by the removal of rainbow trout (*Salmo gairdneri*) from the genus *Salmo* and its installation as *Oncorhynchus mykiss* on the basis of better knowledge of its closest relatives.) Specimen collections provide reassurance that no matter what the name, the animal or plant is still there for inspection and for teaching the next generation of workers.

There is further discussion of data archives in Section 2.3.

iii) Reference sites and reference collections

Reference sites

When it is expected that a change may occur, it is most valuable to be able to compare an area which is likely to be affected with an area which is likely to remain unaffected. This is easier said than done, in a world where the changeability of natural things is normal. In a classic scientific experiment one condition may be altered at a time and the result compared with a 'control' where that condition was not changed. In the laboratory we can have sufficient mechanisms in place to achieve this but in the field a great deal of thought needs to be put into each experiment (for example in testing different treatments of the soil to improve crop yield or in changing the stocking rate of a put-and-take fishery to improve catches in terms of numbers and individual weight of fish caught).

In typical long-term monitoring we are not carrying out experiments but are observing events and trying to associate causes with any changes we see. In the example of a discharge to a river, the sampling site upstream of the discharge may be taken as the reference site. On a global scale it is hard to find sites where the big changes in which we are interested will not make their influence felt. In these cases we may not look for reference *sites*, geographically spaced, but for reference *collections*, spread over time into the past.

Reference collections

Reference collections inform us of how things were at a given place and time in the past. They can be collections of animals and plants, as with museum specimens, or they can be collections of material (air, soil, sediment, water and ice). Each type of collection can help us interpret the present by informing us of the past. For reference collections to have their greatest value, certain rules need to be obeyed. Correct identification (labelling) of the sample is essential and for once-living material the labelling protocol of museums should be adopted.

The usefulness of maintaining a reference *specimen* collection is exemplified by the Beagle and Challenger expeditions of the nineteenth century.

Reference Specimens

The samples collected by Charles Darwin, while aboard the "HMS Beagle" and the first, organised marine biology explorations undertaken aboard the "HMS Challenger", although still used as reference materials, were slated for disposal until it was learned that they were in active use. It was not their historic value but their importance to *today's* research that saved them more than a century later!

Charles Darwin

A special section is given in 2.3 (d) for reference material that is preserved in nature rather than in museums. Such material, some of it thousands of years old, can have a very great influence on our ability to interpret what we see in the present.

Egg-Shell Thinning

Collections of the eggs of peregrine falcons in museums have been used to establish the normal thickness of egg shells in the pre-pesticide era. This highlighted the phenomenon of egg-shell thinning caused by non-lethal effects of bioaccumulation of persistent organo-chlorine residues in adult female peregrines. One of the major reasons for the decline in peregrines was reduced breeding productivity due to egg breakage because of thin shells.

d) Sampling frequency and duration

Sampling frequency means the number of samples taken at one time from one place and the number of sampling visits to that place over a year. In order to make a decision on frequency the project planners have to take into account a number of factors which are probably going to be specific to the programme. Nevertheless some general factors will need consideration. For example:

- the reasons for the monitoring programme;
- the natural variability expected in the material to be examined;
    - *To understand the extent of this variability and relate trends to unnatural events, more samples may be needed in the naturally more variable ecosystems. Even within a single system, such as a river, it is often the case that the communities of sediment-rich parts of the system are less variable than riffle zones. Sampling frequency needs to reflect this.*
- the rate of change anticipated;
    - *Again, the more rapid the expected rate of change, the more frequently samples must be taken, in case an event is missed.*
- biological factors such as growth rates, behaviour and life cycles of the plants and animals within the habitats to be monitored;
- the amount of resource available in terms of manpower, storage and analytical capacity;
- the ability of the sampling site to supply samples without the sampling process itself damaging the ecosystem (see 2.2 (e) below).

Duration: How long a period of sampling is needed for a long-term monitoring record to show its worth? Just as there is no single answer to frequency of sampling, there is no single answer to the length of time over which samples need to be taken. The same considerations apply as those listed under *Frequency*. Obviously, the longer the sampling run the better picture one develops of the conditions at the sample site and whether significant changes are in progress. Ultimately it is statistical validation which determines the minimum period to be covered by the data sets.

At the outset of a long-term monitoring programme, there will not be sufficient knowledge to define this period precisely, so there must be a facility to extend the sampling if the data suggest that this is essential for the purposes of interpretation (see 2.4 (a)).

Sampling must continue for some minimum essential period, depending on the objectives of the programme and the statistical requirement, but the maximum run of data needed may extend to decades or even centuries. After all, civil engineers who are planning coastal defences or the protection of buildings on a flood plain must seek information on conditions that occur perhaps only once over a 100-year period and prepare their designs accordingly, directing where structures might be built and to what standards. Knowing that some events have cycles of recurrence (such as tides and the weather conditions that are influenced by the high pressure zone known as the North Atlantic Oscillation), helps us appreciate that long-term monitoring on the scale of 10 to 20 years can be needed before data become useful in detecting significant deviations from the expected natural rhythm.

> ### Interpreting Extensive Records
>
> In his book 'The Skeptical Environmentalist' Bjørn Lomborg provides summaries of estimates of Northern Hemisphere deviations in temperature over the past 1000 years (Lomborg, 2001). Such a record helps put into perspective the more extreme (and upward) variations in temperature which have been observed since around 1900. For much of the 1000 years there were no direct measurements, although in the world's longest known record (from Central England) thermometers were used from 1659. Interpretations had to be made from ice, tree rings, corals etc laid down at the time. Thus we can examine old data to try to interpret existing data and also to try to determine 'natural' oscillations and variability.

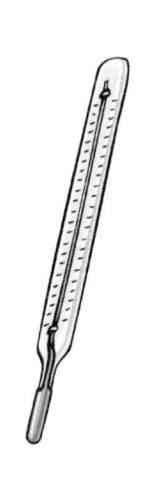

e) Sampling methods

Within our sequence: "Why? – What? – Where? – When? – How?" we now come to the last question: how should samples be taken when they form part of long-term monitoring? Considering the great variety of types of sample: solids, liquids and gases; living and non-living material there is no space in this book to describe an exhaustive list of methods. Indeed there is no need. Some examples will be found in the remaining chapters but the reader is referred to published handbooks on the specific subject of interest. Suffice it to say that the methods chosen must be appropriate to the nature of the sample material and the needs of the programme. The use and availability of data collection devices or instrumentation should be explored to look for what can help to make sampling accurate and easier. Modern electronic systems can and often do provide what is needed at modest cost.

f) Trial periods; sampling limitations; damage from sampling

Before sampling frequency is 'set in stone' a trial period is recommended. There are examples in the literature where a feasibility study has taken three to five years. The advice of a statistician working with experts in the particular scientific disciplines involved is particularly important in planning frequency of sampling. Too often, statisticians are called in to sort out a data set which is inadequate and where proper planning would have prevented or minimised the problem.

As part of the process of co-ordinating the work of the parties involved in the programme the scale of the sampling must be agreed and means found to optimise the value of each bit of information. In other words we should aim for no duplication of effort but to maximise synergy within the programme.

There must be a recognition of the limitations of resources at a sampling site. For instance if a sample of sediment or animal tissue must be collected at each visit there is that much less to sample at the next visit. A properly advanced statistical analysis helps determine the shortest period in which sampling must be performed before it can provide insights on a topic. While the duration of collection is controlled by the specific objectives and the nature of natural variation of the subject, and the 'value' of any individual data point may be small, it is dramatically enhanced by being seen in the context of a long series of such data points.

The amount of data collected should be optimised and related to the quality and desired endpoint and use of the data – more data are not necessarily always better. There are negative points associated with 'over-collection' of data: not just the overloading of scientists but also damage to the sampling sites, especially if samples are taken of species which reproduce relatively slowly. Physical and chemical data can generally be gathered in a non-invasive manner with nothing, or only a very small amount of material, taken away from the site, but with biological material we must consider very carefully the consequences of taking samples away from their habitat.

With all these, perhaps daunting, piece of advice we still have to start somewhere and some day. If a long-term monitoring programme is to be created each piece of information collected adds more than just that point of knowledge. The new data point sits within a field of other points and their relationship to each other has value.

We can summarise this section by saying that in creating a long-term monitoring programme the following points are crucial.

> That planning is thorough;
>
> That key indicator data are identified and collected;
>
> That data collection conforms to an appropriate sampling strategy;
>
> That the sampling process is regularly reviewed in the light of the results obtained.

2.3
THE DATA

In long-term monitoring, as in any form of scientific enquiry, it is crucially important to understand how to interpret data correctly. Major decisions may be taken as the result of the monitoring, affecting not just the ecosystems we hope to protect and sustain but also the livelihoods of people who either earn a living from the natural world or who may be implicated in the causes of any environmental problems made apparent by a programme of long-term monitoring. We must make correct judgements, erring on the side of caution but striking a careful balance between society, economics and environment.

What does the term 'data' mean in relation to long-term monitoring? It simply means facts and figures. These can be

- *quantities* (such as number of individuals of one species, the concentration of a pollutant in water or the number of tin cans on a beach);
- *values* (such as pH, temperature or some assessment of condition set against a scale as in the terms used by the Water Framework Directive 'excellent, good, moderate, poor' for water quality);
- *names* (eg names of species) and
- *any other form* of information from which other facts or ideas may be inferred.

It is useful to **'structure'** data of any kind. This means using some electronic or paper-based system for arranging the data to allow them to be used effectively. For example, a spreadsheet of numbers in columns and rows is a typical and valuable structure. It allows the reader to, literally, 'read-across' to make comparisons or draw inferences, perhaps looking at species lists of rocky shore algae or of temperature records from year to year.

We can describe a set of structured data as a **'database'** or a relational database.

A collection of databases, often held in a computer (but this is not essential) is called a **'databank'**. A long-term monitoring project will normally include one or more databanks, the vital repository of facts and figures from which information is drawn and examined to deliver the findings of the programme. It cannot be emphasised too strongly that data need to be backed up frequently and archived safely.

Even the simplest computers allow *storage* of structured information and rapid *access*, *retrieval* and *analysis* of that information but some global systems are so complex that only large computers have the power to run the lengthy programs of analysis and prediction which may be needed in highly sophisticated studies.

The next sections deal with some of the basic considerations required for even the most straightforward of data sets. The usefulness of the data set will be proportional to:

- the temporal and geographical extent of the sampling;

- the precision of the data;
- the time period in which the data have been collected.

Some of these aspects are discussed below.

a) Sample design: statistical treatment

Designing a monitoring programme is not a simple task and involves a number of steps before you go into the field. These steps are generally the same whatever the scale of the system you are examining.

(1) Decide on the important factors to be measured/monitored that will enable you to find out whether the management goals have been met. This will involve examining existing data and results from either the same areas/species or comparable areas/species, and consulting relevant people or groups in the area.

(2) Consider methods of collecting relevant data and determining the likely quality of those data given your probable resources (in terms of practicality, money and expertise). This may involve carrying out a feasibility study.

(3) Given the quality of data you believe you can obtain, consider how you might analyse those data for trends or sudden shifts: a statistician should be consulted at the outset. Assume various levels of sampling (both in terms of geographic distribution and frequency), and determine the confidence you might have that you would detect changes should they occur. This will usually involve examining statistical confidence given various assumed trends.

From the above, determine the minimum sampling requirements (e.g. in terms of important factors to measure, sampling design in time and space) to be able to address the objectives. Remember that defining a minimum does not mean that you should not aim higher – but it will stop you putting in wasted effort.

It is essential that a long-term monitoring programme should be designed by an ecologist in consultation with managers and stakeholders. Design must allow for statistical analysis. Frequency of sampling contributes to sample size; in statistical analysis this will have a direct effect on confidence in a data point or in the analysis of a trend.

b) Accuracy and precision; quality control

When we measure something, even in the physical sciences, we are aware that our measurement will only be as good as our measuring instrument. With a domestic thermometer we may find that the temperature of a room is, say, 20 °C, but we know that with a better thermometer we might be able to say that it was 20.3 °C. A clinical thermometer designed to determine the temperature inside our mouths is likely to read approximately 98.4 °F or 36.9 °C for a healthy adult but, to be more *accurate* the true value for a particular individual might be 36.87 °C. We also recognise that the temperature of a room will vary from one part to another and that our own body temperatures will vary from hour to hour, even when we are healthy. The same is true for all of the measurements we may need to take during long-term monitoring – measurements will only be as accurate as our method of measurement allows.

In the life sciences we are also particularly aware of the variation from sample to sample, so that the results will cluster around the true value. Whether the cluster is very tight (not much variation between measurements) or more broadly spread we can describe this as the degree of *precision* of the method of measurement. Even when we know that there must be a definite answer, for example the number of species in a given area at a certain time, we know that our counts are only estimates: we may miss something. For other things there will not be a definite, accurate answer. For example, we cannot state that the number of gastropod molluscs in rock pools at a certain height above low tide is x per square metre. We cannot be that precise. We have to say it is *approximately x*, or, more usefully we can make lots of measurements (counts) and say that it is between w and z (usually giving this as a *range (w–z)*), or even, with the use of simple statistics that it is $w \pm s$, where s is a calculated statistic such as the *standard deviation*.

c) Relationships between sampling frequency and intended applications (Ensuring consistency to enable analysis of long-term data sets)

The essential ingredient of a long-term data set is that the data are comparable across time, so that if a change is detected, one can be reasonably sure that it is not the

Tracers helping to decide sampling frequency

It is also possible to use the measurement of one thing to dictate the frequency of sampling for something else. For example, in a river study, we can label a body of water in some way (eg a harmless dye) and increase the rate of sampling at points downstream when the dye reaches each point. This enables us to concentrate on changes within that single body of water – for example the rate of decay of pollutants it may have collected upstream. This overcomes the problem of lateral dispersion of a plug of pollutant as the river moves downstream.

methodology that is responsible for the change. Thus if improvements (e.g. in terms of equipment, methods including sampling design, analytical techniques or personnel) are introduced, sufficient care must be taken to ensure that calibration of any changes is undertaken. One way of doing this is by running methods in parallel for a period of time and comparing the results.

Candidate long-term monitoring programmes should certainly seek to build on pre-existing long-term data sets and to maintain inter-comparability with these data sets.

Where we are looking to spot some sudden change and to pinpoint its onset accurately, frequent sampling may be needed. Where changes are expected to be gradual, a greater period between samples may be appropriate.

d) Who can collect data?

The important principle is that data quality and consistency are maintained by developing and sticking to some standard methods. In this way, all those who collect the data are able to contribute equally. There is great value, where practicable, in involving 'stakeholders' (e.g. local communities, fishermen, managers) in the monitoring process either in helping to collect data (with appropriate training) and/or in regular explanation and discussion of the programme and its results.

e) Quality control

It is extremely important to instil a level of quality control into all aspects of the process, particularly data collection and data analysis - in short the monitors need to be monitored. Detailed protocols must be developed and publicly available and a degree

---

Need for Quality Control

It was salutary in the early days of the acid-rain debate to see the results of one of the trials of measurement of pH across various organisations. They were all sent standard buffered samples of water (which therefore had a definite pH). The results which came back were staggeringly different, giving results whose range was more than one pH unit (a factor of 10). Work was needed to standardise methods and train operators so that the true pH of rain and surface waters could be examined and sensible conclusions drawn.

of audit is carried out to ensure that protocols are being followed. Where laboratory analyses are to be undertaken, appropriate inter-laboratory calibration must be performed. For example, in pollutant studies, international standards have been established and an accreditation scheme established. It is essential that periodic 'blind' tests with reference samples are carried out. In these ways we can control the quality of the data and fine-tune the collection of data if required.

f) Accessibility and other issues

Barriers often exist relating to access to available data, technical compatibility between databases, communications and perceptions of national interest. There may be a lack of common will to provide access to the data. All researchers should be encouraged to provide accessible summaries and reports related to their data within the legal limits (if need be) of commercial confidentiality.

Researchers often have a valid worry about misuse of their data and their right to first publication of analyses resulting from it. Given this they are unwilling to make the data public and often the wider scientific community is unaware of the data until it is published. Fortunately new developments in web-based database technology have to some degree alleviated this problem and it is to be hoped that the situation will improve considerably in the future. The term 'metadata' refers to 'data about data'. The idea of creating web metadatabases is that researchers can make publicly available information about the nature of the data they collect and hold without necessarily releasing the data themselves. Thus a metadatabase may include information on: data collection methodology; type of data; geographical and temporal distribution of the data; quantity of data; and contact persons to apply for use of the data. Examples of these include HMAP (History of Marine Animal Populations - see http://www.hmapcoml.org/?ID=37) and Europhlukes (http://www.europhlukes.net/).

g) Using data sets not originating in monitoring programmes

If we need to establish the existence of an environmental change but don't have a long-term record from our own monitoring, all may not be lost. It may be possible to establish a series of snapshots of the past, which can add up to a useable data set. Here we are observing natural records of the past and analysing this evidence. There are many useful examples:

 i) examinations of fossils to determine past climatic conditions, clarity of seawater, types of sediment;
 ii) using ice cores to extract atmospheric gases from previous millennia (even as many as 300,000 years in Antarctica);
 iii) looking at plankton in lake sediment to establish the previous trophic states of the water (essentially its richness in plant nutrients or organic content) and, for example, the degree of acidification of the lake during past times;
 iv) identifying pollen grains in lake deposits or in peat to give us an idea of

the succession in the surrounding vegetation years ago when these deposits were laid down in sequence;

v) otoliths (inner ear bones) and scales from fish telling us how fast and to what size the animal grew throughout its life (irregularities in the sequence of rings may indicate seasonal changes in food supply or spawning periods);

vi) tree rings, laid down seasonally in temperate zones and, like fish scales, telling us of the rate of growth achieved by the tree each year. A single sample informs us about the individual tree and, by inference, the climatic conditions through the life of that tree. Widely spaced rings indicate good years and narrowly spaced rings indicate bad years of growth. A series of sets of tree rings from different trees, or timbers from trees incorporated in buildings of known date, allows us to piece together the general climatic conditions for hundreds of years into the past. Even a piece of wood which is of unknown date can be used if its pattern of ring width overlaps with that of other pieces of known age. It can itself then be dated and help to put another piece in the jigsaw of data. This science is called *dendrochronology*.

This sort of evidence will not be as precise as long-term monitoring can provide, but is a proxy record where no other is available. It has given very valuable insights into the past.

In addition, and as a very important part of these observations, we have available a series of time checks, (other than the regular annular changes indicated above) set down by natural and unnatural phenomena. Geologists are familiar with major changes in the fossil record where 'discontinuities' have occurred throughout the world, for example the extinction of most dinosaurs 65 million years ago. We have more recent and useful time checks, such as layers of ash in lakes receiving fallout from the eruption of Mount St. Helen's in May, 1980 and radioactive material from the disaster at Chernobyl nuclear power plant in April, 1986. There are also very local benchmarks in time.

Local Benchmarks in Time

When the castle of Frederiksborg was built on an island in a lake near Copenhagen, a small amount of copper from the roof was dissolved and entered the water, becoming incorporated in the lake sediments. This must have occurred between 1600 and 1620, and the distinct layer can still be traced, dating this layer of sediment and helping to date the layers above (younger) and below (older).

2.4
USING THE RESULTS

a) Interpretation – general

We may observe some change in the environment over the course of a few years' monitoring, but we may not be able to determine key aspects of the nature and significance of that change. In terms of its nature, is it

    i) random or directed?

    ii) intermittent or continuous?

    iii) episodic or gradual?

    iv) part of a natural cycle or induced by human activity?

All of these possibilities could be of interest but sorting out those with unnatural origins will require skill in interpretation and probably detailed collective knowledge of the items changing and the possible causes. Natural changes may be part of a downswing that will (or will not) reverse itself naturally over the succeeding years. The problem is to separate these from man-made impacts that may continue and perhaps intensify in the future.

We can use statistical theory to help determine when the variation we observe is greater than is likely to have occurred through chance alone. Part of this analysis will be concerned with understanding the natural variability of the system examined.

Outliers

It is often the case that a small number of results look out of place: they seem either too great or too small to make sense. They are called 'outliers'. It is a matter of judgement whether such outliers should be included in the data set as genuine points or discarded as some kind of anomaly. To aid this judgement the advice of people familiar with the kind of data in question needs to be sought. In addition, it will sometimes be possible to repeat the measurement or observation to see if the same result occurs again. (It is taken as read that if the measurement was made by an instrument this will have been checked against a standard in case its internal system has drifted.) There is always the possibility that the result, although strange, is the first of a series showing some major shift in the system measured. This is where judgement and repeat sampling helps. No data should be thrown out unless there is proof of instrument and/or observer failure.

Weight of evidence

In environmental measurement of any kind, there is always some inherent variability. It can be very helpful to examine several aspects of a situation before reaching a conclusion – so that decisions are made on the weight of evidence rather than any single piece of information.

We must also be on our guard against reaching conclusions on the basis of attractive but false correlations. The environment is so complex that there is always a possibility

> **Inuit Outlier Observation**
>
> A yellow jacket wasp was seen for the first time on Baffin Island recently. The Inuit people who have been there for thousands of years have never seen one before, indeed they don't have an Inuit name for these insects. This could be yet another clear sign of Climate warming but it is an Outlier observation.

that truly unrelated events can seem so mathematically related that they represent cause and effect. Again, experience will help to separate such situations from genuine responses of the environment. It can also be helpful to extrapolate the observation forward or backward to check if it leads to an impossible situation.

b) Monitoring compliance standards

In many ways this is the simplest case for interpretation, although it is not just a matter of stating that any sample falling outside the pre-designated limits dictates that the system has 'failed'. The decision on pass or fail needs to take account of natural variability etc (see section above on Outliers). Often this can be accommodated by benchmarks which state that the situation is in compliance unless x samples per year have failed. The derivation of x is a matter of understanding acceptable and non-acceptable deviations in the system. EU Directives often have indications of this form of judgement. (See WRc: Annual Compliance with Drinking Water Quality Parameters – An Improved Statistical Approach. DETR/DW14757) They may be written in terms of percentages as in the following box.

> **Percentage Compliance**
>
> The European Inland Fisheries Advisory Commission recommended an annual 50 percentile concentration of dissolved oxygen (DO) for salmonoid waters of at least 9 mg/l but said that for 5% of samples the DO could be as low as 5 mg/l.

c) Prediction

The second important role for long-term environmental monitoring is to provide structured information from the present and the past to make predictions about the future. We can consider two general cases: (i) one in which the prediction is directly relevant to the area for which the data are available; and (ii) one in which the prediction is based on data collected outside the area of direct interest.

i) Prediction within the sampled area

Once management goals have been established, it is not uncommon to want to be able to predict the effect of certain external changes (e.g. what will happen if a factory is built nearby). Initially, a long-term monitoring programme can provide a baseline for evaluating whether, subsequently, a change occurs. As the programme continues and data accumulate (for example during and after the building of the factory and during the working life of the factory) it can allow you to obtain sufficient insight into your system to define the extent and direction of change.

Plausible explanations for the observed changes can be sought and may allow you to predict whether the changes are going to produce a significant amount of damage, or, for that matter, a significant amount of improvement. Perhaps as important will be the chance to check predictions and enable the development of better predictive capabilities.

ii) Prediction outside the sampled area

There are plenty of examples of the use of one (or a thousand) case histories to help us predict what would happen 'next time' in more-or-less similar circumstances. For example, we know that boat-ploughing down-slope before planting forests will increase run-off of sediments and cause unwanted events in streams draining the forest. We know the likely consequences of overfishing. Virgil knew the result of letting land lie fallow. The rules established in such cases have a universal applicability, bearing in mind that some environmental systems are more robust (resistant to or tolerant of stressors) than others. Although complexity in an ecosystem would lead one to imagine great resilience, this does not always follow, due to other factors. The Amazonian Rainforest is possibly the most diverse on Earth, but it is terribly vulnerable due to the shallowness of its enriched soils.

An important question, on which we need to assemble data, is whether we can use a totally new set of data to predict consequences for a distant system. Conversely we are now becoming aware of the ways in which global warming can change our local weather. Where predictions are based on 'outside' data, the need for monitoring to ensure whether ones' predictions are correct or whether new management measures must be taken is even greater.

d) Examples (and benefits) of long-term monitoring

The reader is referred to the following contributions to this book, and in general terms to Section 2.1(b), including Table 2.1. We easily identified a series of long-term monitoring programmes that could support the observed reasons for why one would undertake such a study. There are many long-term monitoring surveys from a variety of sources including government agencies, research organisations and private entities that would support and describe past changes and predict probable future events.

Environmental Impact Assessments (EIAs) that help determine the acceptability of development activities are required to characterise the environment in which work will be undertaken and explain any impacts the activities might cause. There is no better source of information regarding the "existing" conditions at a site than a programme operating at the location and designed to collect that very information. All too frequently the data for an EIA are collected in the brief period between site purchase and completion of the construction plans, perhaps taking only months. In one such case, data were collected in mid-summer and an area declared dead. Had it not been for a long-term monitoring programme that showed the area was critical habitat for juvenile fish seeking shelter during the winter months, those species might have been lost.

e) Communication

In any piece of environmental research the work will have little impact unless the communication of the results is carefully considered. For example, we need to use appropriate language for the intended readership, eliminating jargon when we want the messages to be understood widely. It can be of great help to invite a small group of target readers (of whatever interest group) to discuss the needs of their own group. Information may be too much to digest if presented as a single block: it may need to be 'layered' so that the more interested reader can dig deeper into the material to gain more information. Electronic systems (for example a webpage) can be made to provide such a service where the importance of the long-term monitoring project justifies the cost. Websites, brochures, posters, scientific papers and popular articles may all have their place. Another form of data layering uses Geographic Information System (GIS) software. In a GIS, data on a location's characteristics are visually displayed. For instance, a GIS package for Sherkin Island could contain layers with weather, species' use patterns, sea conditions and fishing success. Overlaying the data might reveal how weather influences oceanic conditions and specific species use patterns causing changes in harvests. (See http://www.gis.com/whatisgis/)."

2.5
RECOMMENDATIONS FROM THE WORKSHOP

a) New Programmes

i) Long-term monitoring* of the environment will succeed best if it follows the principles of:
- *Involvement of all interested people** from the outset;*
- *Clear identification of the objectives and benefits (real or predicted);*
- *Openness (transparency) at all times.*

* The term 'monitoring' is to include 'surveillance' throughout.
** We are all shareholders in the Earth.

ii) No matter who conducts monitoring, it should follow Best Practice as outlined in Sections **2.1–2.4** above.

b) Existing or Historic Programmes

i) Data sets which may contribute to long-term monitoring: should be identified and catalogued (eg scope, duration, accessibility) and their existence be made more widely available eg by the European Environment Agency. These data sets may have relationships to each other which have not yet been realised.

ii) We encourage collaboration between the holders of data sets.

c) Special Recommendations

i) Understanding of Biological monitoring requires an understanding of natural history that may well begin at primary school level. We encourage schools to include this concept in their curricula. The role of parents educating their children in environmental matters should not be underestimated, but should be encouraged.

ii) The public should be encouraged to become involved in local monitoring.

iii) Key environmental indicators need to be identified: these may be global, regional or local in scope, depending on the long-term monitoring issue.

iv) In Academia at least, there should be a better linkage between field studies and laboratory work: such integration may help to identify significant and insignificant causes of an observed effect.

v) Businesses should involve themselves in long-term monitoring, reaching out to their local, regional or global communities by
   - *Monitoring within their own area of operations (especially as a condition of a planning consent, but also more altruistically) and*
   - *Facilitating education, co-operation and co-ordination, thus:*
   - *Demonstrating the commitment of Businesses to their customers and their customers' environment.*
   - *Acquiring data on environmental and resource factors which may influence their own ability to continue.*

vi) Predictions need to be checked, as and when possible, and the outcome needs to be reported so that risk reduction measures can be introduced or monitoring programmes tuned.

vii) Technology should complement, and not be a replacement for, observational science.

3.

THE IMPORTANCE OF LONG-TERM MONITORING OF THE ENVIRONMENT

PROCEEDINGS OF THE CONFERENCE ORGANISED
BY SHERKIN ISLAND MARINE STATION

SHERKIN ISLAND, CO. CORK, IRELAND

18TH & 19TH SEPTEMBER, 2003

Editor's Introduction to the Conference Papers

The papers which follow illustrate in detail most of the areas of interest touched on in Section 2. They represent both overviews and detailed analysis of monitoring and its benefits. Some are exclusively from work in Ireland, or Scotland, or North America, or Norway but all can be examined for messages of general applicability to this important subject.

For example:

>We cannot manage what we do not measure.

>If we don't understand the present and are not informed about the past, how can we predict the future?

The reader is encouraged to use this material when considering and responding to the driving forces for establishing high quality long-term monitoring. For example, in Europe, there is the urgent need for every Member State of the European Union to respond to the requirements of the Water Framework Directive in which the duty for monitoring the environment is clearly expressed. This piece of legislation, perhaps above all others, gives Europe the tools for detecting where an ecosystem is under-performing (ie has been degraded). It will allow us to identify correctly the causes of the degradation and then to plan cost-effective measures to repair the damage rather than shooting from the hip and hitting the wrong target! For example, there is little point in looking for sophisticated remedial measures in an area where the requirements of the Urban Waste Water Treatment Directive (European Council, 1991a) have still not been implemented. No amount of monitoring is going to solve a problem whose solution must be civil engineering work, but monitoring will inform society of the depth of the problem to be tackled and the degree of success achieved in its remedy. The return of salmon to the River Thames (one Good News Story that is not covered in this book) and the re-establishment of fish communities in that estuary are the clearest indications of success, and the most readily communicated aspects of long-term monitoring to the general (and voting) public.

It was most encouraging that a Minister of State, Pat the Cope Gallagher, TD, recognised the significance of the event and made time to attend and present the activities and resolutions of his Department.

Indeed the authors are congratulated for bringing their own valued experience and attitudes to help to make this conference a success. No attempt has been made to edit out the personal attributes of the writers. All are commended to your attention.

John Solbé

3.1
OPENING ADDRESS

**Pat the Cope Gallagher, TD
Minister of State at the
Department of the
Environment, Heritage
& Local Government,
Custom House, Dublin 1,
Ireland**

Good morning, everyone.

I'm delighted to be here this morning at the Sherkin Island Marine Station for the opening of the conference on "The Importance of Long-Term Monitoring of the Environment".

At the outset, I want to acknowledge the work of Matt Murphy and his team here at the Station since 1975. The Station produces long-term data and analysis, which are essential to understanding the marine environment around us. Not only have you done that, you have also worked consistently to ensure that the results of the research have been disseminated to a wide audience. Initiatives such as the "Sherkin Comment" and the schools programme have ensured that the public are kept well informed about the ongoing work at the Station.

The attendance today, and the level of participation in the preparatory workshop, are, I feel, a reflection of the esteem with which the Station, you Matt and your colleagues are held in the environment community, both nationally and further afield.

I should also take this opportunity to congratulate again the community here on Sherkin Island on winning last week the Island Award in this year's Tidy Towns Competition. The commitment of the island community and its many visitors to maintaining its overall environmental quality is an example to us all.

The role of long-term environmental monitoring

Turning to the conference theme, it is essential that we continue to develop strong environmental policies and that environmental considerations are integrated into sectoral policies. Environmental monitoring has a major role in this task. The policies that lead environmental change, and the conviction among people that change is necessary, both rely in substantial measure on good monitoring and research, as well as on dissemination of the results.

Many changes in the environment occur gradually and may, of course, be masked by natural fluctuations. So it is necessary to establish at an early stage monitoring systems which are capable of detecting and assisting in the understanding of these changes so that we can prevent problems, rather than just deal with the consequences. But the usefulness of long-term monitoring systems is very much dependent on the comparability of the data produced over the measurement period. Thus, where such systems are being set up, close attention needs to be given to the methodologies proposed for making the observations to ensure that they will deliver accurate data on a consistent basis.

Looking at the Irish environment today in broad terms, the EPA *Millennium Report* and the OECD *Environmental Performance Review of Ireland* have both acknowledged that the overall quality of our environment remains good. The same broad conclusion was reached in the European Environment Agency's report for the Environment for Europe Conference published before the summer.

While we will have a fresh assessment next year in a new EPA State of the Environment report, drawing on the results of a wide range of monitoring activities, at this stage it is clear that there are no grounds for complacency. Like elsewhere in Europe, Ireland's environment is now under pressure – from the nature and pace of economic activity, together with the associated patterns of consumption.

We must be clear. Damage to the environment and depletion of natural resources can only be avoided by resolute measures to protect the environment and make our development considerably more eco-efficient.

I would like to comment briefly on a number of the critical areas and the role of monitoring in addressing the issues.

Water Quality

Water quality has been monitored in Ireland for a longer period than most other environmental media. In fact, the same biological assessment method has been used to monitor water quality in rivers since 1971.

The EPA Water Quality report, for the period 1998–2000, has found that, for the first time since national reports were prepared in the early 1970s, there has been a reversal of deterioration in the river system. Two thirds of the length of Irish rivers and 93% of lake waters are now classed as unpolluted. Though significant problems remain to be addressed, including contamination of groundwater and the tendency to eutrophication in tidal waters, compared to other EU waters Irish waters are generally in good condition.

We need to keep it that way and improve further. So since 1997 we have pursued a comprehensive, integrated strategy to tackle all sources of eutrophication which very much anticipated the requirements of the EU Water Framework Directive's whole river basin approach. The Directive puts a strong emphasis on the monitoring of flora and fauna in aquatic ecosystems on a long-term basis. It requires a co-ordinated approach to water management in respect of whole river basins with a view to maintaining the high status of waters where it exists, preventing deterioration in the existing status of waters and achieving at least "good status" in relation to all waters by 2015.

To meet these requirements, we are actively promoting the establishment by local authorities of comprehensive river basin management projects for all waters.

Air Quality

In relation to air, the demand for information on pollutants has increased in recent years, due to greater awareness of the role of such emissions in a variety of major environmental issues such as climate change, acid deposition and smog. On-going monitoring of emissions is crucial in order to provide information on which to base action to avoid damage to human health and the environment. This is one area where the need for environmental knowledge knows no boundaries, as some of the pollutants are truly global, with evidence, for example, of the impacts of human activity to be found even in the most remote parts of Antarctica.

Irish inventories of atmospheric emissions data, which have been developed by the EPA, have been broadened to include more substances and have become more complete over the last 10 years. These data have been integrated with those from other countries to facilitate the compilation of global data sets, which are essential to understanding change on a global level.

At a local level, too, such data are important. Most local authority air quality monitoring focuses on smoke and sulphur dioxide in urban areas. For example, 30 years of air quality data are now available for Dublin City, covering pollutants such as sulphur dioxide, lead and carbon monoxide and these allow the development of appropriate policy responses. Such monitoring also enables the public to be informed if emission levels reach concentrations that would present health risks across wide areas, especially for sensitive populations.

Climate change

As you all know, climate change is recognised as one of the most threatening global environmental problems and this is supported by the scientific evidence. The rate of change is more rapid than previously projected with new and stronger evidence that most of the warming observed during the last 50 years is attributable to human activities. Potential impacts of climate change include rising sea levels, changing rainfall patterns, increases in temperature and weather disturbances.

A key source of data on climate change in Ireland is the network of meteorological monitoring stations supported by Met Éireann, which is now part of my Department. At 14 locations throughout Ireland, observations of temperature, precipitation, wind, sunshine, cloud and other aspects are made on an hourly basis, with many records extending from the 1940s. Long-term monitoring of rainfall and temperature began in

Ireland in the 19th century, and some stations – such as those at Valentia Observatory and Birr – have continued right up to the present.

Similarly, records of the start of growing seasons made in some of Ireland's major public gardens, such as the National Botanic Gardens in Dublin and the John F. Kennedy Arboretum in Co. Wexford, provide useful long-term data on climate trends. While these types of secondary data may originally have been compiled for other, quite different, purposes, they are now being used to give us information on changes in our climate and point the way to potential future changes.

These long-term data sets were critical to the preparation of a report last year by researchers in Maynooth and Trinity College on climate change indicators for Ireland, which shows that we in Ireland are mirroring, albeit at a somewhat delayed rate, the trends in climate change apparent at the global level.

Research and monitoring of this nature provide a basis for real long-term planning by the agencies concerned, including planning for adaptation to climate change.

In terms of greenhouse gas emissions, meeting our commitment under the Kyoto Protocol is a core challenge for sustainable development policy in Ireland. Realising our Kyoto emissions target of 13% over 1990 levels is one of the most demanding tasks that we as a country face. Accordingly, we have initiated a review of the *National Climate Change Strategy* to ensure more focused and intensive implementation.

Conclusion

To conclude, I have no doubt that the next two days will contribute to better understanding of the environment and of the importance of long-term monitoring. The conference will promote greater awareness of the value of our environment and the care it demands and deserves from us all.

In addition to the work, I hope you will all find time to sample the pleasures of the island and of island life. I wish you an enjoyable and productive conference.

Thank you.

3.2

Is Short-Term Monitoring Sufficient?

G. E. (Tony) Fogg
Prof Emeritus in Marine Biology,
School of Ocean Sciences,
University of Wales, Bangor,
UK

THIS is a question which scientists of the mid-nineteenth century may well have pondered on.

One of the most famous and influential scientists at that time was Baron Alexander von Humboldt. A man of considerable wealth, he went below his social status to become a scientist, an expert on geomagnetism, geology and botany and a pioneer in climatology, oceanography and biochemistry. He travelled widely in South America. He saw nature as a whole, that is, his approach was holistic and his studies extended over wide areas and long periods of time. One of his achievements was to establish a network of magnetic observatories co-ordinated to make observations at specified times over long periods. This led to important advances both in the understanding of the earth's magnetism and its practical applications in navigation. Unfortunately for Humboldt his work became overshadowed by the precision, beauty and commercial value of the burst of experimental work carried out by scientists such as Faraday. Their success with short-term reductionist science swamped the holistic approach and those who wanted to get on confined themselves to the laboratory and neglected the environment. They put aside the fact that whereas replicable results can be obtained under conditions of adjusted and controlled temperature, pressure and concentrations, out in the environment these factors may be variable and different with the particular process being studied enmeshed in a complex web of other processes, so that the outcome might be quite different. Even biologists in those times seemed to have preferred to be in the laboratory to dissect organisms and fit them into Darwin's theory of evolution. Only agriculturalists and marine and freshwater scientists had perforce to carry out experiments and observations in the natural environment, often for long periods.

One of the most famous examples of long-term monitoring is that started in 1843 by Laws and Gilbert in the Broadbalk wheat field of what is now the Rothamsted Research Institute in Hertfordshire, UK. The aim was to establish the sources of the nutrients required in plant growth. Plots were treated with different mineral nutrients or manure and yields of wheat grain and straw, as well as exhaustion of nutrients and effects of weather, were recorded. Rothamsted had some pioneer statisticians who, using among other things the Broadbalk data, introduced valuable new methods into the treatment of agricultural data in the 1930s. The Broadbalk experiment still continues and the main results are that yields from plots supplied with artificial fertilizers and those that had farmyard manure are the same and that below the level of 21 inches of rain per year yield becomes correlated with rainfall rather than sunshine.

Outstanding long-term monitoring in the aquatic environment has been carried out at the Windermere Laboratory of the Freshwater Biological Association. This was initiated in 1945 by Dr J.W.G. Lund, who concentrated on a diatom, *Asterionella*, abundant in the English Lake District, accumulating detailed records of changes in its numbers and factors likely to affect its growth. Eventually, with the help of other workers, data extending over half a century have become available. This has not only provided a model for other studies on plankton but has supplied invaluable information which has enabled water management to be carried out with greater efficiency. Data like this, obtained for a specific purpose, may be coupled with other data from separate contemporaneous studies of different phenomena with profitable results. Dr Glen George has used Lund's data together with some from other lakes, including Lough Leane, only forty miles from Sherkin, together with meteorological records to examine the effects of regional-scale variations in atmospheric circulation on plankton. He has shown that the quasi-cyclic weather events influence the flux of nutrients and thus phytoplankton dynamics. The feature known as the North Atlantic Oscillation has the most marked effect on production in the lakes.

Turning to the sea, the most extensive long-term monitoring is that based on continuous plankton records obtained with a sampler designed by the late Sir Alister Hardy and first used in the Southern Ocean in 1926. The recorder is towed behind a ship, filtering out plankton as it moves through the water onto a moving strip of silk gauze. These recorders have been towed systematically by merchant ships and ocean weather ships in the North Atlantic. This began in 1932 and still continues giving information, valuable particularly for fisheries, on the distribution of plankton and fish eggs. More modest marine monitoring has been carried out on shores. An example of this is the survey carried out by Dr Eifion Jones (who has tragically died since the Sherkin Workshop) and his colleagues on the shores of Anglesey over a period of years. Semiquantitative observations of intertidal flora and fauna were made monthly and subjected to multivariate analysis. It was found that seasonal changes, affecting almost all the shore life, varied considerably from year to year. Variation between shores, even if close to each other, could be considerable and the structure of communities appeared to be by no means entirely explicable in terms of the floristic and faunistic patterns in the text books. Here the inadequacy of short-term surveys is very evident. Long-term monitoring can be expensive, often monotonous and occasionally a waste of time. If one is interested in changes in the past these disadvantages may not arise – the miscellaneous particles and materials trapped in successive layers in marine and freshwater deposits, peat bogs and ice provide records which can often be precisely dated and can yield an amazing variety of information covering thousands of years. Pollen, diatom frustules, remains of zooplankton and chemical constituents in its bottom deposits can give detailed pictures of a lake's history and this in turn can be related to changes in climate, geological catastrophe and human activities. Air bubbles trapped in polar ice can tell us how air temperatures have varied with concentrations of greenhouse gases in the atmosphere and indicate what the course of global warming may be in the future. Present changes are not so easily and precisely monitored. One attempt to short-cut is by mathematical modelling. Suppose, for example, one needs to predict the future growth and behaviour of marine plankton in a given sea area over a period of years. The growth rates and responses to temperature, water movements, nutrient supply, light intensity and its variations, predation and so forth can be determined in the laboratory and expressed in mathematical equations. These can be assembled in a model and environmental factors such as temperatures, tides, turbulence, daily and seasonal variations in light can be fed in and the computer predicts plankton behaviour under the expected conditions. This usually works reasonably well, the model producing graphs showing changes with time which match approximately records obtained directly from the environment. One such mathematical simulation was constructed in great detail for Narragansett Bay (Rhode Island, U.S.A.) and seemed satisfactory until it was tested for other regions. Adjusting annual cycles of solar radiation and temperatures to those found 19° South produced violent and meaningless oscillations in the graphs.

I will finish with a tale of long-term monitoring with which I had some connection. In 1973 there began to be worries that gaseous discharges (e.g. chlorofluorocarbons) from human activities might cause destruction of ozone in the upper atmosphere and lead to penetration of damaging amounts of ultraviolet radiation to ground level. Systematic monitoring of ozone had already been started in two of the British Antarctic Survey (BAS) stations in 1957. In 1979 BAS, like many other scientific institutions, became desperately short of money and it became necessary to charter research vessels as cargo ships and to consider closing down one or more of the Antarctic stations. Sir Vivian Fuchs, when he was Director, had set up a Scientific Advisory Committee and the problems were duly referred to it. There was one representative of the Natural Environment Research Council (NERC) and he, quite reasonably, pointed out that the atmospheric chemistry unit of BAS had found very little change over 22 years and was not doing any really exciting work. The U.S.A. had recently sent up a satellite to make measurements of concentrations of various substances, including ozone, in the upper atmosphere and were willing to pass on the data to BAS. Therefore the atmospheric chemistry unit of BAS was superfluous and could be closed down, the financial savings solving BAS's problems. Some members of the Committee were reluctant to lose scientists whom they knew to be of first class quality. Argument went on for two hours or so until I, as Chairman, felt we had to put the matter to the vote. This came out as four in favour and four against. As Chairman I had a casting vote and, fearing greatly that I was bankrupting BAS, I decided that we should keep the unit. Three things happened after this;

- The Falklands War broke out and BAS was able to give the Government so much help with information on weather and ice movements that Mrs Thatcher gave it an extra £9 million a year. We were no longer insolvent.
- The BAS scientists at Halley Station in the far south, continuing to make ozone measurements with their antiquated equipment, made the startling discovery in the spring of 1984 of the 'ozone hole'. The U.S. satellite had apparently missed it. It had registered some 50 million ozone determinations but the scientists were at a loss to know where to look amongst all this for something interesting. Going over the raw data again they were chagrined to find the ozone hole there as large as life. The right people had been on the right spot at the right time and BAS long-term monitoring had been completely vindicated.
- NERC decided that it was not going to tolerate this sort of thing and disbanded the Scientific Advisory Committee replacing it with a new committee made up of members appointed by headquarters.

Is short-term monitoring sufficient?

The answer to the question is no.

__Tony Fogg__ has been particularly fortunate – long holidays on his grandparent's farm, family friends and teachers who were keen naturalists, at university distinguished botanists, ecologist and plant physiologists as professors and increasing links with the Antarctic where he eventually stood on Observation Hill by the cross in memory of Scott and his companions and looked down the route they had followed to the South Pole. Added to all this was a wife who loved walking in the country and going to the theatre. He has managed to write a large number of scientific papers, which he says "have now, no doubt, found their way to rubbish dumps; and eight books on plankton, photosynthesis and polar matters, which had excellent reviews but didn't make my fortune"!

__Editor's Note:__ Since the Sherkin Island Marine Station Workshop/Conference, we are sad to announce the death of Tony Fogg, following a short illness. He will be missed by all his friends and colleagues.

3.3
AN OVERVIEW OF THE SOCIAL VALUE OF LONG-TERM MONITORING

John F. Solbé
Environmental Consultant,
Dol Hyfryd, The Roe,
St Asaph, Denbighshire,
Wales, LL17 0HY, UK

Introduction

In this chapter I have briefly summarised some of the important elements of programmes of observation of the environment, which we term environmental monitoring or surveillance. I have written this from the perspective of the questions Society may raise in our perceived need to know something about the environments in which we live. Some of these questions are given below.

Surely, as individual members of Society, our primary considerations are:
- Am I safe?
- Is my family safe?
- Is the ecosystem safe now and will it remain safe in the long term?

But there will also be secondary considerations such as:
- Are the improvements to the environment and my family's safety, promised by a regulatory authority, really appearing?
- Have we now recovered from some recent example of environmental damage such as an oil spill?

We need to consider what tools we have to help answer those questions. For example we may have regular health checks. We may be given safety instructions when intending to use a household chemical such as an adhesive or bleach. These instructions will have been based on an assessment of the intrinsic properties of the chemical such as flammability or toxicity and an assessment of our possible exposure to such hazards. Together they form a 'risk assessment'.

These are all of great value, but under the subject of 'Long-Term Monitoring' the tools to be discussed in detail are those of professional monitoring and professional surveillance. However, we should not forget that personal awareness of our surroundings, which may combine both monitoring and surveillance, can be of a high standard and based on sound knowledge and good principles. (See for example, the subject of 'phenology', below.)

What is the difference between surveillance and monitoring?

We need briefly to discuss the difference between these two terms but 'monitoring' can be used as a sufficient overall term for the rest of this Chapter. They have been clearly defined by Hellawell (1978).

> *Surveillance* is a continued programme of surveys systematically undertaken to provide a series of observations in time.
>
> *Monitoring* is surveillance undertaken to ensure that previously formulated standards are being met.

In other words, when we just keep an eye on the environment, making careful observations of course but not setting the result against some standard requirement, this is surveillance. Examples might be recording the date of the first swallow to arrive each Spring, the species list at a rocky shore site, the number of whales in a given area. We might hope for certain results but we do not have in mind a standard result which we would regard as a 'pass' or 'fail'. Much of environmental 'monitoring' is therefore, strictly speaking, surveillance. Figures 1 and 2 give contrasting example, in which the same data are used, but either set or not set against some standard framework. Figure 2 (standard framework) is therefore an example of monitoring while Figure 1 shows a

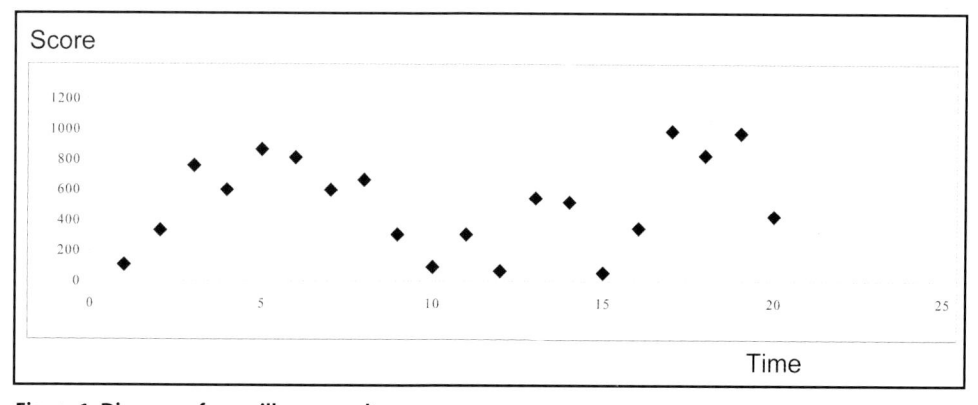

Figure 1: Diagram of surveillance results

> These allow us to see the state of the environment, particularly looking out for good and bad trends, without having any particular result in mind and without necessarily knowing possible causes or effects.

result of surveillance, a recording of data without a preconceived standard. The monitoring data in Figure 2 could be, for example, the number of coliform bacteria in the sea off a bathing beach, where we have a medically derived risk assessment informing us that above a certain number of such bacteria there is likely to be a risk to health.

Of course, a failure within a monitoring data set could be some low number (such as too low a concentration of dissolved oxygen, indicating some form of organic pollution, or too low a pH, indicating an acid discharge of some kind) instead of a high number, as in the case of faecal coliform bacteria.

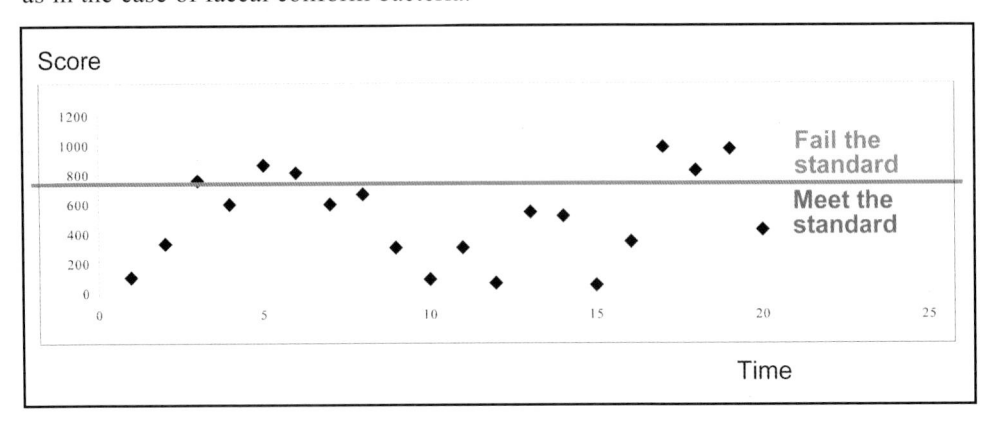

Figure 2: Monitoring results

Here the data are compared with a preconceived numerical standard and we can check whether any sample is within or outside compliance.

In either case one of the principal driving forces nowadays for monitoring or surveillance is to support our aspirations for 'Sustainable Development' a relatively recent way of looking at ourselves and our environment bringing together three important focal points: Social Progress, Economic Growth and Ecological Protection. Each of these can be monitored, although it is only the last which is the focus of long-term monitoring as discussed in this book.

Examples of the value of surveillance

It may be useful to remind ourselves where surveillance has revealed previously unsuspected problems in the past.

- *Acid rain* – Possibly as early as 1850s changes began to occur in lake diatoms, toxicity to fish and food organisms in areas of hard geology such as the west of Scotland, Wales and the granitic regions of southern Scandinavia. These largely remained unnoticed until the 1960s (but we know some of them happened a century before because our modern-day studies of ancient lake sediments show us the changes from sensitive to tolerant forms of algae over the years). Later, as information began to accumulate that something was wrong, it was the surveillance work of fisheries officers, hydrologists and meteorologists which demonstrated the extent of the problem. The causes were identified and included the burning of large amounts of fossil fuels for power generation resulting in the deposition of rainfall which was made acidic by dissolved oxides of nitrogen and sulphur.

- *Loss of song birds due to pesticides* – In 1962 Rachel Carson published her now famous book 'Silent Spring' in which she described the loss of songbirds in her university campus in the USA. Surveillance here was at first just an observant person who noticed a change. The reasons for the change are now well understood but at the time of her book scientists were only just becoming aware of the fact that certain chemicals can accumulate in living tissue and can be passed on from prey to predator increasing in concentration *biomagnifying* as they passed along the *food chain* (or more typically the more complex *food web*).

- *Changes in the sexual characteristics of animals* – In the 1980s scientists studying dog-whelks began to notice changes in the reproductive organs of these animals. This discovery was the result of general surveillance work, not a response to predicted problem, but it became one of the first-observed examples of a widespread phenomenon of disruption of the natural functioning of hormones in various types of animal ranging from oysters to alligators to fish and seagulls and Man. In dog whelks and oysters the cause was a chemical (tributyl tin) used in anti-fouling paints for boats and large ships. In alligators and gulls it was various chlorine-based chemicals with several uses including pest control. In freshwater fish however, where the anglers and fisheries scientists were first to spot the problem, 'sex hormone mimicry' is now known to be principally caused (in UK at least) by natural human female hormones and their analogues used in birth control and hormone replacement therapy. This last group of chemicals reach rivers having passed unchanged through sewage treatment works.

- *Climate change (Global warming)* – Now that we know that climate change is occurring more rapidly than in the past few millennia we can begin to follow changes such as the raising of sea levels (10cm since 1900), as atmospheric warming, due to increases in atmospheric gases, particularly carbon dioxide leads to warmer seas, melting of Polar ice and reduced density of sea water. Surveillance follows the trends: there is no particular standard for us to monitor against, unless we choose to compare the present with some ideal period in the past.

- *Destruction of ozone layer* – The Antarctic observations described by Tony Fogg are not an example of surveillance but a result of theoretical work which predicted the problem followed by monitoring of the ozone layer, which eventually became a monitoring of the size and position of the hole.

How else can both surveillance and monitoring help us?

Knowledge and measurement of the environment over extended periods can help us understand better any of the areas of current environmental concern. These are, in descending order of importance

    i) Ecosystem destruction or replacement

    ii) Over-exploitation

    iii) Eutrophication

iv) Climatic disruption

v) Pollution other than eutrophication

As an example, the protection of waters, from any of the problems outlined in (i) to (v) above, will benefit enormously from environmental monitoring and indeed may not be possible without it. We must make opportunities to monitor rainfall, climatic changes, the abstraction, storage and distribution of water, our fisheries, the invertebrate animals that inhabit our streams, lakes, estuaries and seas, the quality and quantity of water around us and measures of pollution either by added nutrients (eutrophication) or by toxic or smothering substances. If we don't measure, we can't manage.

So, we know why and where we want to monitor, but what do we need to consider when designing a sampling programme? The answer may be summed up as all those aspects which make the programme 'fit for purpose'. What does this mean? It means a programme in which there is a clear definition of the

- Objectives;
- Locations of sampling sites (targeted habitat types, sites of special interest etc);
- Duration of the programme (sufficient to answer the questions posed);
- Frequency (set at such a level that important events are captured and not missed);
- Methods, clear and designed to achieve the objectives.

The purposes which can be considered as suitable subjects for environmental measurement include

- *Identifying an ecosystem which is 'healthy' or of High Ecological Status according to the requirements of the Water Framework Directive*

    Throughout the European Union the need under this Directive is to establish which parts of a river basin are already of a satisfactory standard ecologically and which need to be brought up to standard. In both cases the situation has to be monitored, in the former case to ensure that the good situation is maintained; in the latter to follow the responses of the ecosystem to any remedial work. The Water Framework Directive is a major undertaking of the EU and will take many years to be brought fully into operation. It is open-ended and thus a perfect model of a need for long-term monitoring.

- *Defining ecosystem responses to man-made chemicals in the presence of all other influences on ecosystems (eco-epidemiology)*

    Ecosystems are rarely unchanging. As well as responding to the seasons and weather they show the result of many complex interactions among and between species and habitats. This makes it difficult to separate what is a natural phenomenon from a man-made one. Long-term monitoring has a double role to play: it demonstrates change, but by a careful consideration of the timing and nature of the change, combined with other studies, it can be used to define causes.

- *Defining biologically significant changes from a wealth of impermanent environmental changes (adverse v non-adverse effects)*
- *Predicting future environmental problems against timescales (in essence defining how close we are to the 'cliff edge' – the point where an ecosystem may have almost used up its assimilation capacity and be about to show a major downturn in some attribute such as biodiversity);*
- *Predicting the response of ecosystems to mixtures of xenobiotic chemicals;*
- *Extrapolating ecotoxicological understanding between ecosystems;*
- *Valuing ecosystem services (such as the ability to consume wastes and render them harmless, the provision of fisheries for recreation, the provision of aesthetically pleasing landscape) and using this in socio-economic analyses of regulatory measures.*

Are there good examples of environmental monitoring?

- Sherkin Island Marine Station monitoring of plankton in the Atlantic around the Fastnet and of rocky shore monitoring on the island;
- Continuous Plankton Recordings in the Atlantic;
- Freshwater Biological Association monitoring of fish populations and environmental conditions in Windermere;
- Fishery statistics collected nationally, and intensively (R. Bush, Corrib system);
- Deep-frozen samples of fish eg in Finland;
- Meteorological records throughout Great Britain and Ireland;
- *Amateur phenology.*

What can we do to help?

- Encourage environmental education;
- Exhort authorities;
- Alert authorities of problems (anglers have a particularly useful role);
- Join phenology initiatives.

Phenology is the study of the timing of natural biological events each year, such as the arrival and departure of birds, first and last sightings of butterflies, bud-burst of trees etc.

It provides a long-term (decades) record of phenomena linked principally to climate change.

In the UK, phenology is organised by
www.woodland-trust.org.uk/phenology

Concluding thoughts:

- Long-term data series are of enormous value, even when we are not aware of their future uses.
- We do need to CONNECT chemistry, biology, land-use, hydrology, climate…if we are to understand what our monitoring and surveillance data are telling us.
- We must build our confidence by ensuring that we have taken sufficient samples, preferably using standardised (but at least well-described) methods; EU / CEN / ISO standards are the ideal.
- We must try to quantify and understand natural variability so that we can separate natural changes from man-induced ones.
- Mapping systems (Geographic Information Systems) can be of great value when they are used to build up a picture of current situations and trends. *For example, the significance of acid rain is well demonstrated by overlaying rainfall, pH of rain and surface geology to identify sensitive areas.*
- The EU should encourage the education and retention of experts in environmental monitoring.

To repeat the questions I posed at the beginning:

Am I safe?

Is my family safe?

Is the ecosystem safe now and will it be in the long term?

Are the promised improvements appearing?

Have we now recovered from recent damage?

And to remind ourselves of the importance of long-term monitoring:

I can't manage if I don't measure.

If I don't know what happened before, how can I guess what will happen next?

John Solbé *was Head of the Environment Group of Unilever's Safety & Environmental Assurance Centre from 1988 until his retirement in 2001. He was trained as a freshwater ecologist in the University of Wales, graduating in1963. He immediately joined the UK Government's Water Pollution Research Laboratory (WPRL) and remained there (and in its descendant Laboratory, WRc) for 25 years, working on a wide range of environmental problems. These included treatment of sewage, the effects of pollution on freshwater fish and new chemicals legislation within the EU. In Unilever he continued his interest in science-based approaches to risk assessment both nationally and internationally and maintains active links with the subject as a consultant and as a Visiting Professor at the Universities of Cranfield and of Newcastle upon Tyne.*

3.4
LONG-TERM MONITORING
– A MEDIA VIEWPOINT –

Alex Kirby
Environmental Journalist,
28 Prince Edward's Road,
Lewes,
East Sussex BN7 1BE,
UK

"You should not seek to bribe or twist
the honest British journalist.
But seeing what the man will do
unbribed, there's little reason to."

MOST of us can worry about only a few things at a time. Professor Robert Worcester, who heads the British polling organisation Mori, made the point with what he calls the Alligator Principle. He said: "Few of us worry very much about global warming when we're up to our neck in alligators. But global warming is a greater threat to most of us than alligators are ever going to be." We can bear only so much reality, and that usually means we'll worry more over what's immediate than about what is – or at least seems – remote. So long-term monitoring of anything has got the dice loaded against it from the start. To worry about alligators when they're all around you is only rational, after all.

We recognise that monitoring is important because we recognise the existence of risk. It's rational to try to protect ourselves against the risks we perceive we may face. But few of us are entirely rational in the way we judge the relative importance of the risks we face. In 1996 the UK Department of Health planned a rating system for certain risks (it was originally going to be associated with new drugs and with some medical procedures, but was broadened to include other categories as well). The system stated that there was a high risk, defined as odds on dying in any one year of one in a hundred, linked to certain hereditary diseases. Moderate risk, with odds between one in a hundred and one in a thousand, was associated with steady smoking. Odds of a thousand to ten-thousand, a low risk, went with road travel, while travel by rail and air was classed as very low risk, with odds from ten to a hundred-thousand. Some possible accidents were classed as being of negligible risk, with odds of dying put at less than one in a million – being struck by lightning, for instance, or catching Creutzfeldt-Jacob Disease through eating beef from a mad cow. The odds of winning the British lottery were also described as negligible.

The Department of Health was entirely rational in its approach. But it reckoned without those of us (perhaps the majority) who continue to think it's worth buying a lottery ticket every week. We know in our rational minds that it's a lot more dangerous to drive a car from central London to Heathrow airport than it is to let a complete stranger fly us on from there to Cork, or wherever. But we still terrify ourselves with improbable nightmares about what an air crash would be like, while we drive to and fro without a care. Every day in Britain people die horrific deaths in traffic accidents – and we all agree it's dreadful. But we don't stop driving. Every few years there's a horrific train accident, and we call for limitless spending on rail safety, and try to avoid ever again getting on a train at all. And then we have a smoke to calm our nerves.

Being irrational is part of the human condition, and so long as we recognise that's how we often operate there's no great harm done. It's as if we are resigned to accepting quite high odds that a small, personal tragedy will sooner or later probably engulf us, while we refuse to accept very long odds on a grotesque and improbable contingency.

What has all this to do with long-term monitoring, though – if anything? Only this – that long-term monitoring is laborious, painstaking, often dull but always a rational way of taking the pulse of our world.

We need to monitor what we're doing to our world, and what it may do to us. In July, BBC News Online carried a report headlined "Drought warning for Ireland". The first sentence read: "Ireland's climate is likely to change dramatically in future years if nothing is done to change its environmental policy, a report has warned." The report was from the Environmental Protection Agency, established to monitor Ireland's environment. If climate change really is a greater threat to most of us than alligators ever will be, we have to thank the people who've helped us to discern this emerging threat – the climatologists, the scientists from a range of disciplines, who over the long term have been monitoring the effect on the Earth of the build-up of carbon dioxide and the other greenhouse gases.

Before most of us knew the climate as a whole was changing, we'd done some serious worrying about the ozone hole, and that's thanks to the work of still more scientists, including Joe Farman and his colleagues from the British Antarctic Survey, who discovered the damage that chlorofluorocarbons and other chemicals were doing to the ozone layer. And the fact that they were allowed to continue their long-term monitoring and discover the hole at all is thanks essentially to that intuitive casting vote by Tony Fogg.There are many other examples of the way in which the monitors have time and again saved the day by blowing the whistle to alert us to an emergent threat. The case for long-term monitoring is pretty rock solid.

And it's not hard to think of possible problems where monitoring could provide a definitive answer if it were allowed to. One of those is the use of depleted uranium, or DU for short, in military weapons. In its solid form, DU is mildly radioactive, and 1.7 times as dense as lead. It's a very effective material for anti-tank weapons: when a DU round punches through a hard target it disintegrates into a spray of burning dust, which everyone acknowledges can damage your health if you inhale it. The western allies used it in the first war against Iraq, in 1991, and some coalition members used it in this year's war. Many Iraqis believe DU is at least partly responsible for what they say is the increased incidence of cancers and child abnormalities since 1991. But the US and the UK say the reports of higher rates of cancer and birth defects are statistically unproven, because there were no reliable statistics before the war and because there has been no reliable health monitoring since then. You might say the Iraqis would say that anyway. But it isn't just the Iraqis. Significant numbers of American and British ex-service men and women have reported a range of complaints since 1991 – Gulf War Syndrome, as it's known. But they're having great difficulty in proving their case – and they say that's partly because there's been no systematic long-term monitoring of their condition. Many of them believe they are being left to die off one by one until there is nobody left to monitor, and no case left to answer. And DU has been used in the Balkans, and perhaps in Afghanistan. Long-term monitoring would help the Gulf veterans and the Iraqis to prove that they are ill, or not, and to know why. But there wasn't any.

Another example, geographically closer to home, is the Beaufort Dyke, a deep undersea trench in the North Channel between Scotland and Ireland. I heard in the mid-1990s that it had been used for dumping chemical weapons, so I asked the UK Ministry of Defence. It told me the Dyke had been used as a dump for chemical munitions since soon after the end of World War Two, and that about 100,000 tonnes had been jettisoned there. We filmed people on the coasts of Ireland, Scotland and the Isle of Man who showed us phosphorus grenades that had been washing up on their shores. A year later, the Ministry told me the Dyke had been in use as a dump since shortly after World War One; that the munitions jettisoned there amounted to not 100,000 tonnes but more like one million; that not everything had been dumped in the Dyke itself, as some cargoes had been thrown overboard in shallow water near the coast; and that in any case there was no record of what had been put into the Dyke. So much for long-term monitoring.

So it's rational to say we should be doing more long-term monitoring across the board. But there are several problems. One is that monitoring can give us the data we need. But it's still up to us to decide the significance of the data, and whether it is in fact significant at all.

Some years ago I was lucky enough to meet a man who's one of the most original thinkers about the environment in Britain today, Professor James Lovelock. He's probably best known for propounding what he calls the Gaia Hypothesis, which says, roughly, that the Earth is a huge organism, and humanity little more than a pest species or a parasite, which the Earth will destroy when our activities become too damaging to the whole organism. But what I remember from the time I talked to him is something quite different. He said, if not exactly in so many words, that our ability to measure pollution – to monitor it – was now so advanced that we could detect substances in amounts so minute that they would have no discernible effect on us or on other life-forms. I sometimes think of Jim Lovelock when I hear yet another demand for an improvement in the quality of the UK's water supply, to take one example. We have allowed ourselves to be persuaded that the water which comes out of our taps is suspect. So we spend a massive amount on bottled water, despite evidence that it's often of poorer quality than the tap water. If people want to waste their money like this, that's their business. But the single-minded preoccupation with eliminating every last trace of pollutant from the public supply makes a sad contrast with the plight of communities which lack a proper public supply at all, and with the thousands who die daily from waterborne disease and poor sanitation or none at all. If you want a bigger bang for your buck in adding to the sum of human happiness, forget domestic water quality and spend the money making sure the wretched of the Earth at least have something they can drink. So often, when we try to make environmental improvements, we end up allowing the best to be the enemy of the good. We've become so obsessed with finding cutting-edge, state-of-the-art, hi-tech solutions to every problem that we've forgotten how to bodge, to make do and mend, to settle for what's good enough. And what's just about good enough can make all the difference between life and death. So we shouldn't allow ourselves to over-interpret the evidence our measuring supplies us with. It's long-term monitoring that will help us to see whether the measurements show a trend which should lead us to act, or whether what we have is in effect a snap-shot.

Another problem we have to face is that the choice is not always between two perfectly-matched goods: it may be a question of deciding which is the lesser of two evils. Being concerned to improve the environment is often a messy, pragmatic business, which doesn't leave a lot of room for absolutism. What long-term monitoring can do is show us the existence of a trend: that can save us from knee-jerk reactions. It can help us to choose the lesser evil.

There are some areas of life where the level of proof people expect from monitoring may take too long to acquire. One such case is probably genetically-modified crops – GMs. To prove that they will neither harm human health nor have an adverse effect on wildlife would take a lot of work, and many years to complete. So should we press ahead regardless, trusting in the benefits the bio-tech industry says we can look forward to, or tread very carefully, making the precautionary principle our touchstone?

I don't know. My own feeling is that, on scientific grounds alone, GMs may provide some answers in particular situations. I also feel that the pure science is being forced to take a back seat, and that the reputations and profits of members of the bio-tech industry are driving society pell-mell towards decisions it needs quite a lot longer to resolve.

That leads to a further question – whether we should allow the absence of conclusive data from long-term monitoring of a particular environmental technique or innovation to stand in the way of progress towards its introduction. Again, I don't know. Again, I think it's a question of horses for courses, of deciding every case on its merits. The absence of evidence of harm is by no means the same as evidence of its absence. But the opposite is true too: we may not be able to prove something is beneficial, but all the same we should not dismiss the possibility that it may be – and that it may be better than the existing options. In an ideal world we might well say no to GMs. In a world where much of Europe's farmland is a huge outdoor factory, drenched in chemicals and bereft of wildlife, we may want to think a little longer.

I am a fan of long-term monitoring. I think it's alerted us to all sorts of perils we'd otherwise have ignored. But let's remember one qualification. Many of the horrors of recent years have come from developments that weren't monitored at all. Who'd have guessed that feeding the diseased remains of ruminants to their living relatives would have given us mad cow disease? And who bothered to monitor what was happening anyway? Who'd have thought a night shift trying to see what would happen if you switched off a nuclear plant's safety systems would trigger the world's worst nuclear accident? Who'd have thought anyone would leave open the bow doors on an English Channel ferry – and that several inches of water would have turned it over as a result? But all these entirely unpredictable accidents happened. Life's full of surprises, and a lot of them aren't nice ones. Sod's Law dictates that we'll be caught out by the horror we haven't thought of, not the one we've monitored for years.

But the clincher for long-term monitoring must be what it does to reinforce memory. Today we are becoming increasingly short-termists, perhaps because of three things: a four or five year political cycle, an economy that depends on novelty and obsolescence to maintain productivity, and a range of media that are increasingly losing the plot. If I were a cynic, I'd say in countries like the United Kingdom we have too many media chasing too little news. So non-news, anti-news, becomes a headline. We are losing our grasp of history, and of geography. There was the recent story, perhaps not apocryphal, of a sub-editor recruited to a London tabloid, who on his first day in the newsroom asked a colleague: "What's our house style? That country in the Middle East – do we call it Iran or Iraq?"

And there was Gus, the managing editor of the TV station in the sit. com. 'Drop the Dead Donkey' some years ago. He was a delightful man, but without a single thought in his head, apart from a desire to get viewers for his station. The fictional Gus is said to have been modelled on a former senior colleague of mine who I'll call X. X was a kindly man who helped me a lot, and I remember him fondly. I also remember one of his underlings saying to me once: "Alex, you really do have to respect X – he is genuinely as shallow as he seems."

I remember a delegate at a previous Sherkin conference, who came out with one of the most memorable definitions I've heard. He said: "The environment? That's what we do to where we live." So it is. And long-term monitoring is the way we shall learn how what we do to where we live is gradually altering it, for better or worse.

Alex Kirby is a British journalist and broadcaster. He covers the environment for BBC News Online and presents BBC Radio Four's environment programme Costing the Earth. He works with charities, NGSs and other groups to enhance their media skills. He is a member of his local wildlife trust in Sussex in southern England.

3.5
THE ROLE OF GEOLOGY IN LONG-TERM MONITORING

Peadar McArdle
Director,
Geological Survey of Ireland,
Beggars Bush,
Haddington Road, Dublin 4,
Ireland

Introduction

We in Ireland do not suffer the effects of the major disasters which are most commonly associated with geology – volcanoes and earthquakes. So is there really a role for geology in long-term monitoring? I believe that there is and that this role is already being partly discharged, either here in Ireland or elsewhere in Europe. It is important that it not be discharged in isolation: a multidisciplinary approach is essential with common standards underpinning what should be web-enabled Geographic Information Systems (GIS), reporting data in real time.

It is important to identify the priority issues where geology might make a contribution. In reporting on insurance losses from natural disasters in 2002, the Economist (24th May 2003, p.102) noted that the biggest losses were due to extreme weather events: floods in Europe, spring storms in America and hurricanes in the Caribbean. All relatively commonplace, and geology can contribute to how we monitor, understand, forecast and remediate these types of events. While geology is popularly associated with rare extreme events – such as the potentially huge tsunami that

Figure 1: Earthquake damage in Kobe, Japan.
Source: The Economist, 24 May 2003.

might be generated by coastal cliff collapse in the Canaries, as reported a few years ago – these are very rare events compared with others which are much more likely to impact on all our lives. It is these that we should focus on.

Safeguarding our communities

Extreme weather events in recent years have given rise to extensive *flooding* including in areas which have no historical records of flooding. Much effort is devoted to engineering solutions for remediation and prevention. However it would be unwise to ignore the historical baseline provided by flood plain sediments. These were mapped in detail in the nineteenth century by the Geological Survey of Ireland (GSI) and reflect areas flooded throughout our varied climatic history since the melting of the ice sheets, say 15,000 years ago. While we now realise that the past may not be a good guide to the future, nor past extreme weather events a guide to future patterns, the distribution of flood plain sediment here reflects the maximum areas flooded since the Ice Age. Their extent in the Cork area is greater than that known to have been flooded in the past, but does include recently flooded areas. The important contribution of geology lies, not only in providing this baseline study, but in establishing a monitoring grid to predict the occurrence of landslides and soil erosion, two aspects which were crucial recently in Central European floods.

Protection against pollution is a major issue, especially in urban areas and wherever land usage has changed. This points to the need for excellent integrated data storage in

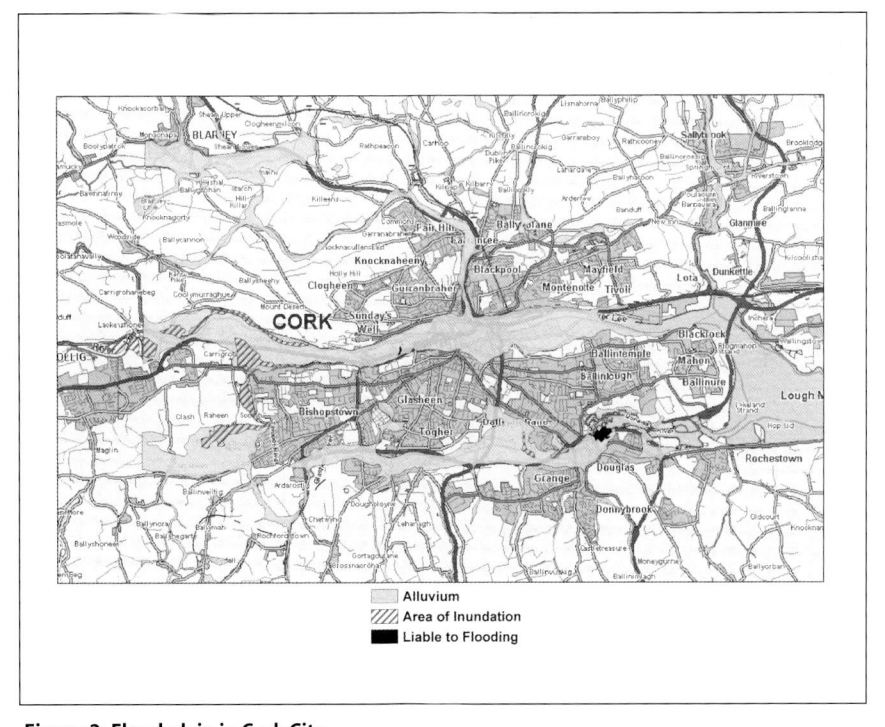

Figure 2: Flood plain in Cork City.

a GIS, where it can easily be accessed in the future. Sites of former industries or mines may become available for new uses. One example is the old mining areas such as Silvermines where studies of contaminated land are being coordinated by the Environmental Protection Agency (EPA). Geochemical baseline surveys are available for this area and can be used for future monitoring (Figure 3a). Another form of pollution can be gaseous. Methane gas can escape from covered landfill and abandoned underground coalmines, while radon gas may infiltrate into foundations from underground cavities, especially karstic ones (Figure 3b).

With increasing urbanisation and balanced regional development, there is an increasing focus on *secure transport systems*. Geology can contribute by minimising impacts from ground conditions. The DART[1] was recently disrupted by the landslides along the South Dublin

Figure 3a: The variation of lead content in soils at Silvermines, a valuable baseline for monitoring this contaminated land in the future. (Source: EPA)

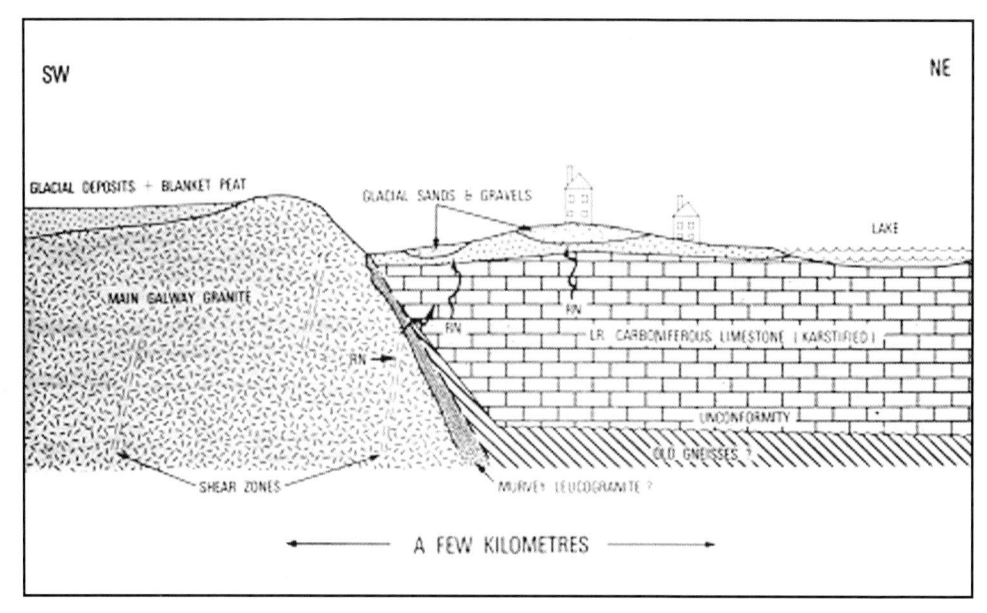

Figure 3b: This shows a cross section through the Moycullen area, County Galway, where granite sourced radon has migrated into, and concentrated in, cavities in nearby limestones, posing a potential risk for overlying households. (Source: GSI)

---

[1] Dublin Area Rapid Transport

coastline. U.K. cities are locally undermined by coal and other mining and there is a famous photo of a bus protruding from a suddenly developed cavity in a suburban street. North Dublin residents are currently concerned with the potential for subsidence associated with underground routes such as the Dublin Port Tunnel. Satellite-based radar surveillance has been routinely gathered for the past decade and provides a rich and high-resolution baseline database for future monitoring of ground movement. Indeed by providing historical data, it can also be used for monitoring of current activities. With a vertical resolution of less than 1cm, it is possible to track the incidence and intensity of subsidence or swelling of the land surface (Figure 4).

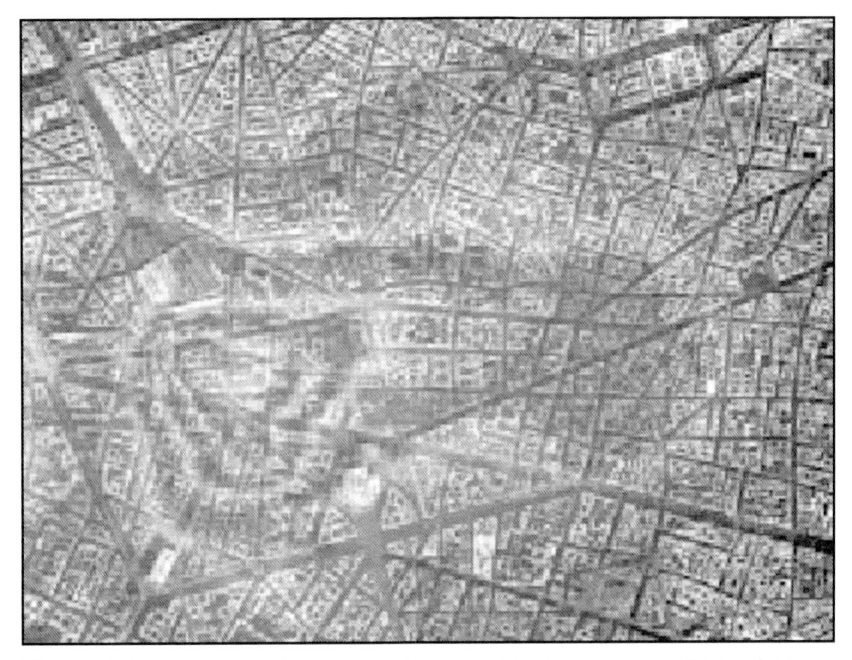

Figure 4: When underground water pumping ceased in the area around St. Lazaire in Paris, the water table rose and gave rise to ground swelling, each ring representing a 3mm increment of uplift.

(Source: GMES Terrafirma Project)

Protecting our Environment

A comprehensive series of studies on *water resources* is currently underway in response to the EU Water Framework Directive and designed to ensure the reliability and quality of water supplies into the future. These studies are managed by agencies such as the EPA and Office of Public Works (OPW), as well as the local authorities, and GSI has contributed to those concerned with groundwater. These studies will represent a remarkable set of baselines by which future developments can be gauged. In the case of groundwater, GSI contributes through Groundwater Protection Schemes and Source Protection studies that it undertakes collaboratively with local authorities and others.

There are two areas where geology can make further contributions to long term monitoring. One concerns suspected point sources of pollution. Aerial surveying techniques, in this case electromagnetic, can be carried out quickly and cost-effectively in such circumstances. The technique images the conductivity of the subsurface material. While earlier images might show that leachate (as distinct from pure non-conductive water) was confined to the limits of a particular landfill, later images could outline any leachate plume extending away from the landfill. This provides an early warning of potential pollution and gives accurate information should remedial action be contemplated. A recent survey of waste dumps in a UK coalfield area indicated the presence of such a plume of polluted groundwater extending away from the dumps (information from The British Geological Survey, see Figure 5). Without such an airborne survey the first indication of pollution might be its appearance in water supplies.

Groundwater monitoring is the statutory responsibility of EPA and it undertakes this on a national basis with the support of various other bodies, such as OPW, local authorities and GSI. It involves monitoring the groundwater quality and the groundwater level and is currently carried out on a satisfactory basis. However geology might support the monitoring more effectively: it may at present be difficult to be sure what the sampling or measurements represent. Many monitoring sites may not be well characterised in geological terms – there may be no adequate well logs or inadequate information on their surrounding areas (e.g. flow directions and gradients, permeability, vulnerability). This is in no way a criticism of EPA or any other staff: there is simply a huge amount of geological work to be completed at perhaps more than 300 sites (such as illustrated in Figure 6) and this needs to be planned and resourced in the future, as a contribution to effective monitoring in the longer term.

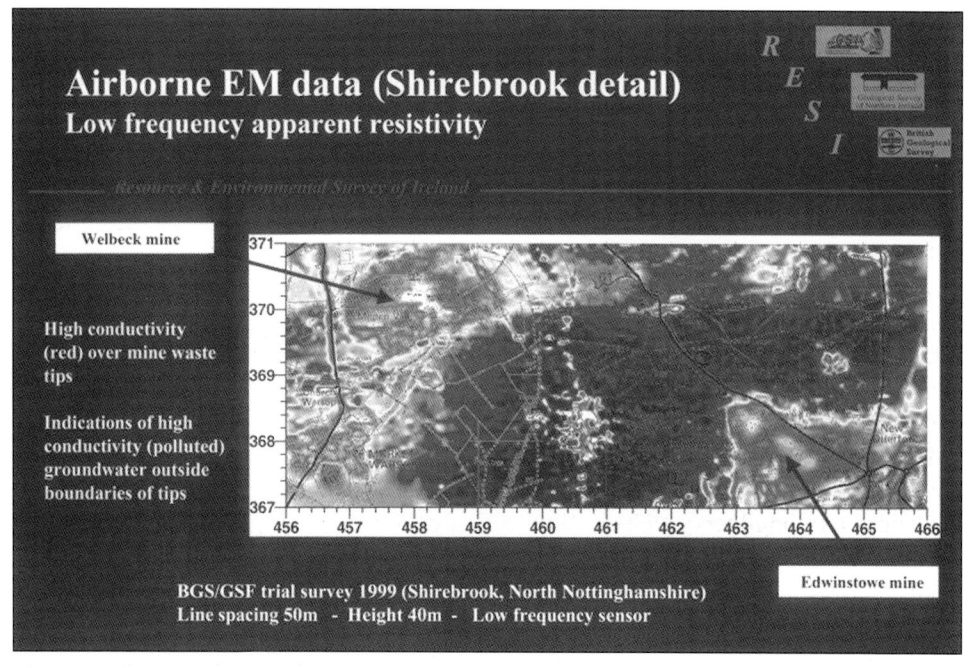

Figure 5: Airborne EM data (Shirebrook details) - UK coalfield area.

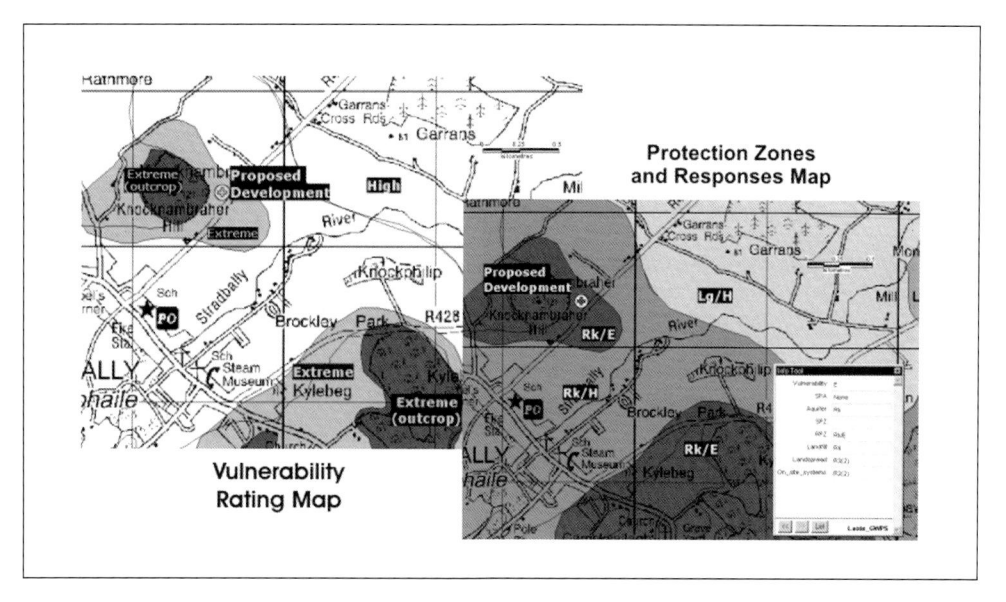

Figure 6: Groundwater Source Protection Study and Ground Water Vulnerability Mapping.

The monitoring of *food quality* is a matter of great concern to us all at present. The application of geological techniques can assist at the agricultural phase of food production. For example the excessive application of fertilizers, which led to the publicised destruction of vegetable crops in north County Dublin some years ago, can be monitored by aerial surveying involving potassium spectrometry, a technique that is commonly available on survey aircraft. Another example could be caesium contamination of upland grasslands arising from the Chernobyl disaster in the 1980s. Caesium has a relatively short half life which means that it decays fairly rapidly. The important thing for grazing sheep is that they spend adequate time on low-caesium grassland before entering the food chain. High resolution images can help farmers to manage the process effectively and cheaply. The pattern is so detailed that most farms will have low-value fields.

Our *upland areas* are among the most attractive and valued of our landscapes and yet their vulnerability is not widely appreciated. Creating significant baselines and establishing monitoring patterns can be assisted by techniques such as remote sensing and hyperspectral scanning commonly used and interpreted by geologists. Uplands are one type of landscape where dramatic human influences can be traced. Soil erosion by hill walkers or over-grazing has been a major factor. In a country that has actually recorded one fatality by avalanche, we are more prone to difficulties caused by landslides and bogflows, such as those which occurred in Mayo in 2003 (Figure 7). Biodiversity is another long-term issue in these areas, especially where forestry is planned. Aerial electromagnetic techniques can be used to map changes in soil permeability in a semi-quantitative manner. In Finland it has been found that tree species are sensitive to levels of permeability and species selection may be based on the result of such techniques.

Figure 7: Views of the landslide which occurred near Pollatomish, County Mayo in 2003.

Understanding our climate

The reality of climate change is now almost universally accepted. We are less concerned here with the drivers of this and the extent to which they are influenced by anthropogenic factors. We have already seen some impacts of extreme weather events, giving rise to flooding for example. Let us consider a couple of additional practical examples. Global warming should lead to a *rising sea level* and to a shift in shoreline positions. Many suspect this may already be happening but there is no adequate database to test whether this is so. Shifts in coastal sedimentation patterns as well as changes in river behaviour (both erosional and depositional) will be additional influences resulting from global warming. At present we are obliged to rely on seabed charts which may be up to a century out of date. As part of the National Seabed Survey, which is managed by GSI, in partnership with the Marine Institute, we have undertaken some airborne laser surveys of coastal areas, such as this image of part of Clew Bay and the approaches to Westport Harbour (Figure 8). Other agencies have gathered similar data elsewhere around the coastline. Primarily aimed at improving navigational accuracy and enhanced maritime safety, these high resolution surveys will serve as important baselines for the future, when repeat surveys can be undertaken. Once more these are quick and cost-effective. As far as vertical resolution is concerned, the occurrence of sand waves and a seabed cable may be distinguished in the Clew Bay example. Such images will have additional significance as far as long-term monitoring is concerned. For example it is important to identify weak cliffs so that the incidence of landslides and coastal erosion can be predicted. In addition these features will help to predict the extent of land inundation, especially in times of extreme weather events. The

Figure 8: Lidar (Light Detecton and Ranging) image of the offshore area of Clew Bay.

extent of inundation will be related not only to the extent of sea level rise and the other factors already noted; it will also depend on the energy and direction of major storms.

Moving offshore, climate change may have a dramatic impact *on our seabed and its resources*. Firstly oceanic warming may lead to changes in the distribution of fish habitats. Through multidisciplinary collaboration we are beginning to understand the detailed seabed controls on our fisheries which should lead to more efficient fishing and an improved seabed environment. A time series of sound velocity profiles through the water column (Figure 9) can be used to interpret the effects of global warming on oceanic circulation and topography. At ambient seabed temperatures, hydrates are frozen, but even very slight increases in seawater temperature may lead to their unfreezing in potentially huge amounts. Hydrates tend to be concentrated on continental slopes and can release globally significant amounts of methane and other gases – which are greenhouse gases. We have reason to believe some exist offshore Ireland. Indeed some major undersea landslides, in themselves hazards for fisheries and offshore

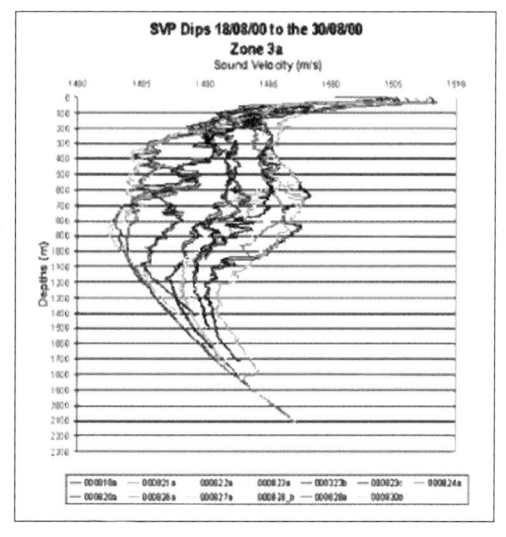

Figure 9. Sound Velocity Profile

petroleum installations, may well be indicators of unfreezing hydrates. Repeat ship-based surveys, targeted on areas such as these may be very important in monitoring the seabed environment into the future. But we are not limited to such surveys. Moving beyond the technology involved in the exploration of the wreck of the 'Titanic', multi-disciplinary seafloor monitoring networks are now being established in various parts of the world to detect important seafloor events and report them in real time. In due course, Ireland needs to participate in such monitoring. In time it may even assist in remediating the current high levels of greenhouse gases by identifying suitable undersea spaces where carbon dioxide may be sequestrated. This is an area of active research elsewhere in Europe but the potential in Ireland where we are facing real challenges in meeting our Kyoto obligations, has yet to be evaluated. Can we afford to continue to ignore this aspect or the potential hydrates problem?

Communicating our results

There is little value in long-term monitoring if its results are not communicated effectively to all stakeholders in a timely manner. In practical terms, such a system must be web-based, but few organisations have reached that stage yet. The provision of early warnings can save both lives and money. For example, The Economist recently reported that early warnings concerning El Niño in 1997–98 saved California an estimated $1

Protection against pollution
- Contaminated/hazardous land (geochemical and gas surveys)

Transport routes
- Subsidence monitoring (Satellite radar)

Groundwater resources
- Point sources of pollution (airborne EM)
- Understanding the monitoring (hydrogeology)

Food quality
- Grassland caesium, fertilizer application (airborne spectrometry)

Conserving uplands
- Soil erosion, landslides, bogflows (remote sensing)
- Suitable for forestry (airborne EM)

Climate change
- Sealevel and shoreline position (airborne laser)
- Seabed resources/hazards (ship-based multibeam, seafloor monitoring networks)

Communicating results
- Web-enabled GIS

Figure 10: A summary of these techniques and their application in long-term monitoring

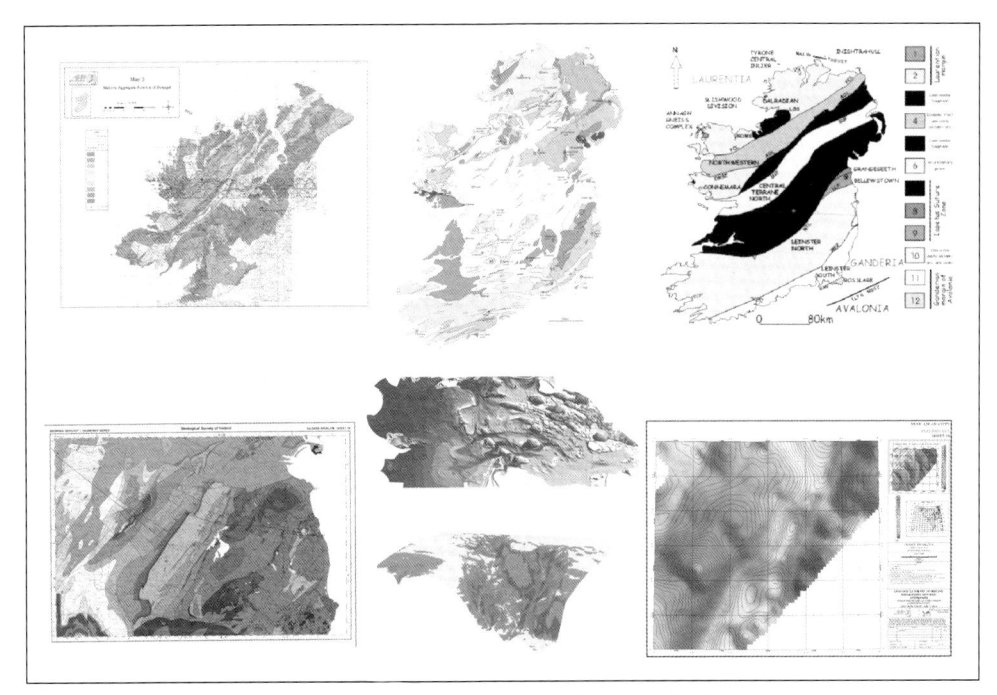

Figure 11: A range of images of different aspects of the Geology of Ireland.

billion, while the rest of USA annually saves $200–300 million on the basis of smaller-scale warnings.

However we also need to invest nationally in a range of survey techniques to which geology has much to offer by way of understanding and interpretation.

Geology can contribute to long term monitoring, but its contribution will be effective only when it is undertaken in a multidisciplinary fashion, with results integrated in a digital geographic information system (GIS) which uses common standards. All our databases, wherever they are based, must be compatible with each other, driven by common standards. Baseline data are essential for effective future monitoring and many of these already exist. The bedrock of the country may be persistent in many of its qualities, yet our interpretation and understanding of it will change according as society's needs change.

So who needs Geology? We all do – but it must be integrated with other disciplines to deliver an accessible user-friendly product. We must operate in a digital environment and be productive in communicating our message. We have a powerful toolbox – but the community is not yet aware of its benefits.

Peadar McArdle is Director of the Geological Survey of Ireland is based in Dublin. He graduated in geology from University College Dublin, where he was awarded a PhD. After periods in Africa and the Irish mining sector he joined GSI. GSI is the National Geological Agency.

3.6
THIRTY YEARS MONITORING WATERS, WEEDS AND FISHES

W.S.Trevor Champ
M.F. O'Grady, P. Gargan,
P. Fitzmaurice and P. Green,
Central Fisheries Board,
Unit 4, Swords Business
Camp, Balheary Road,
Swords, Co. Dublin,
Ireland

Introduction

The Central and Regional Fisheries Boards were set up by Statute in 1980 to replace the Inland Fisheries Trust (IFT) and seventeen Fisheries Conservancy Boards; they are charged with the protection, management and development of Fisheries.

The midland lakes, Loughs Sheelin, Ennell, Derravaragh and Owel and several of the large "western lakes", Loughs Carra, Corrib, Mask, Conn/Cullin, Arrow and Inchiquin are all managed as brown trout fisheries (Figure 1). These alkaline waters contained fast growing brown trout with an average weight of approximately 1kg (Kennedy and Fitzmaurice, 1971). These trout stocks were recognised as a significant natural exploitable resource and development work commenced on Lough Sheelin in 1951. This was then extended to the other lakes by the mid 1960s. The midland lakes at that time contained clear water characteristic of alkaline lakes of this type. In February of 1971 a significant blue green algal bloom (*Anabaena* sp)

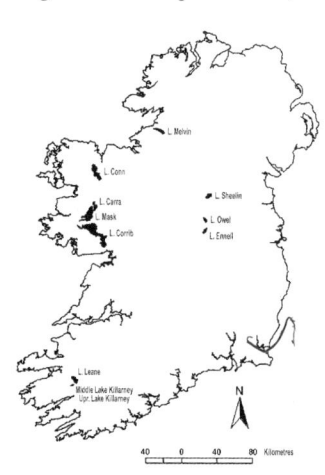

Figure 1

developed on Lough Sheelin which clearly signalled a change in the environmental conditions of the waterbody. This was a new phenomenon in Ireland never having been recorded on these lakes previously. The development was recognised as a symptom of eutrophication which presented a serious threat to the trout angling industry, then an important national amenity and significant earner of tourism revenue.

Eutrophication Assessment and Impact

An assessment programme was initiated in 1971 to establish the trophic status of the midland trout lakes and following a significant and dramatic change in the ecological conditions in Lough Ennell between 1972 and 1974, the programme was extended to large western lakes in 1975 (Champ, 1977). The CFB with RFB assistance have continued this lake water monitoring programme. Comprehensive surveys of all the lakes was beyond the resources of the Boards and a monthly midlake sampling programme was initiated to monitor total phosphorus, water transparency (clarity) and algal biomass as indicated by chlorophyll extract, the key indicators of eutrophication (Vollenweider, 1971 and OECD, 1982). Sampling is confined to the open water station in each lake or basin. It is acknowledged that the programme is restricted but, because these Irish lakes in general are exposed to the prevailing winds and therefore well mixed, the open lake sites were chosen to give a meaningful indication of the overall condition of the lakes (maximum return for minimum outlay). This proved to be the case in Lough Sheelin (Figure 2) and Ennell which underwent dramatic changes in trophic status over the 30 year period (Champ, 1979; Dodd & Champ, 1983; Foy, et al., 1996). These lakes demonstrated significant ecological changes, e.g. expansion and contraction of the lake bed area colonised by submerged vegetation (Figure 3) which were clearly associated with open water characteristics (Champ, 1993). The frequency, intensity and duration of blue green algal blooms increased and the trout stocks were observed to decline as lake conditions deteriorated (became excessively enriched or hypertrophic). Evidence of the stock decline on Ennell was compiled from anglers catches, trout encounters in gill nets and redd counts in spawning streams (Table 1). The vegetation and trout stocks recovered in Lough Ennell (Figure 4) when that lake was restored to mesotrophic status for an extended period.

The adequacy of the open lake monitoring programme as an indicator of overall lake

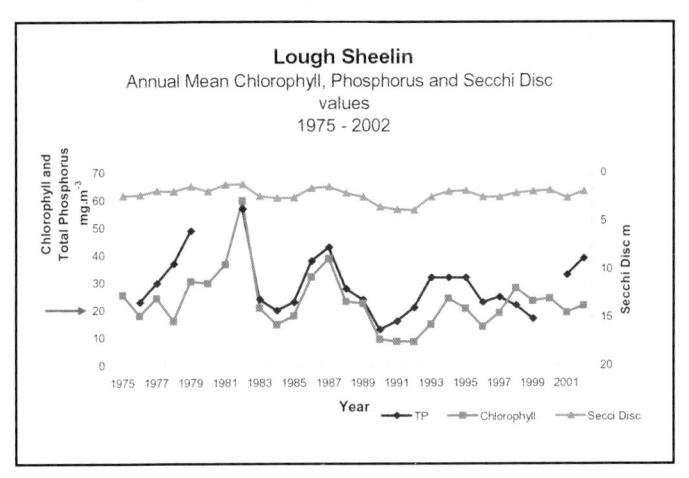

Figure 2

Table 1: Evidence of declining trout stocks in Lough Ennell

	Method	1968	1969	1970	1971	1972	1973	1974	1975	1976
Gill nets for pike	No. of trout reported	829	1,657	1,427	835	439	471	464	211	295
	No. per net	6.3	9.6	8.7	7.5	3.9	2.1	2.8	1.8	1.4
Spawning Trout	Redd Counts	1,963	2,641	3,271	2,422	1,891	1,512	1,144	260	138
Creel Census	No. of trout reported	4,434	4,097	2,064	3,222	2,033	790	161	159	32

Source: Champ 1979

Figure 3: Distribution of charophytes in Lough Sheelin 1972–1991

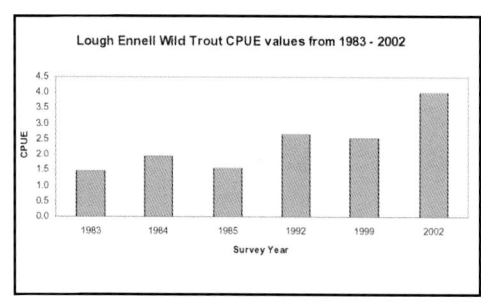

Source: Champ 1993

Figure 4 *(Source: Champ, 1998)*

Lough Ennell Wild Trout CPUE values from 1983 - 2002

condition has been brought into question in some of the western lakes which are only moderately enriched (mesotrophic), e.g. Corrib/Mask where the eutrophication process has progressed more gradually (Figure 5) and to a much lesser degree than in Sheelin or Ennell (Champ, 1998). In Lough Corrib the mean annual open lake chlorophyll though still only indicative of mesotrophic (or moderately enriched) conditions has increased from about 2 to 5 mg.m^{-3} in the period 1975 to 1980 to about 3 to 7 mg.m^{-3} in recent years (Figure 6). This change, although modest relative to some other Irish lakes (e.g. Sheelin), is significant in the context of Lough Corrib (upper basin). In both Corrib and Conn (mesotrophic and mesotrophic/eutrophic categories respectively) the char (*Salvelinus alpinus*) population (a salmonid species sensitive to pollution) which had populated the lakes since the Ice Age (Champ, 1977) has disappeared. The cause is thought to be eutrophication related, significant accumulations of decomposing blue green algae (cyanobacteria) were observed (Figure 7) intermittently in these lakes in the 1980s when the char disappeared from both water bodies (McGarrigle *et al.*, 1993; Champ, 1998). A substantial increase in nutrient loading – 18 to 35 tonnes molybdate reactive phosphate per annum has been demonstrated to have occurred in Lough Conn between 1980 and 1992 (McGarrigle *et al.*, 1993).

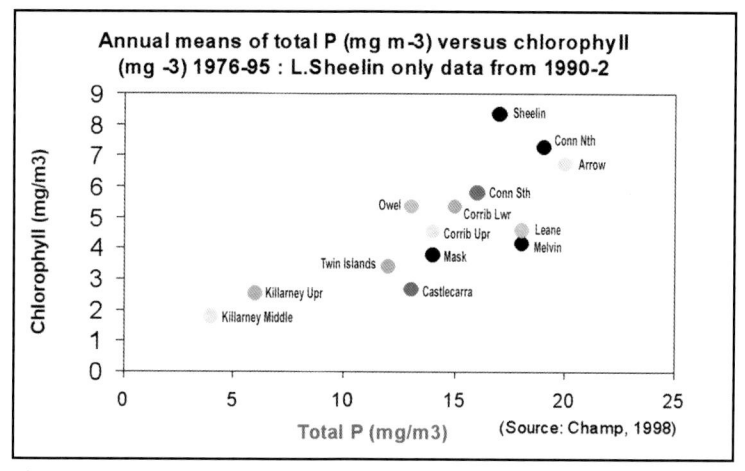

Figure 5

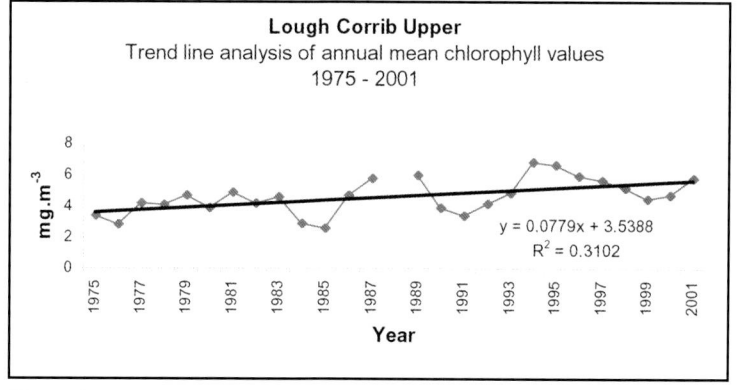

Figure 6

Figure 7: Significant accumulations of decomposing blue green algae were observed intermittently in Lough Conn and Lough Corrib in the 1980s.

Monitoring trout in Lakes

The importance of monitoring, freshwater fish stocks, has been clearly illustrated by the work of the Central Fisheries Board over the last thirty years. In the late 1970s a technique involving the use of experimental gill nets was designed to monitor the structure and relative density of fish stocks in Irish lakes (O'Grady, 1981; O'Grady, 1983). The value of using this technique is illustrated here in relation to fish stock monitoring data series from Loughs Sheelin and Conn.

Virtual annual monitoring of the trout populations in Lough Sheelin from 1978 to 2001 illustrates major changes in the relative density of the trout stock over time which can be related to the trophic status of the lough (Figure 8A). Data compiled at less frequent intervals for Lough Conn illustrates similar major changes over time (1978 to 2001). The Conn data series also shows how rapidly fish stock densities can change (Figure 8B) – rudd (*Scardinius erythrophthalmus*) were known to be present in Lough Conn in small numbers at least since the 1960s. However none were captured in annual monitoring exercises until 1998. Yet by 2001 a very large stock of rudd was present. Roach (*Rutilus rutilus*) introduced to Lough Conn in the 1990s have also expanded rapidly – the rudd/roach ratio in Lough Conn in 2001 was 1.0/1.49 (O'Grady and Delanty, 2001). All of these fish stock data were measured as Catch per Unit of Effort Values (CPUE). The individual CPUE values for all of these studies were generated by dividing the number of fish captured by the netting effort undertaken it shows the CPUE for trout in Lough Conn declined from 6 to 1 between 1984 and 2003.

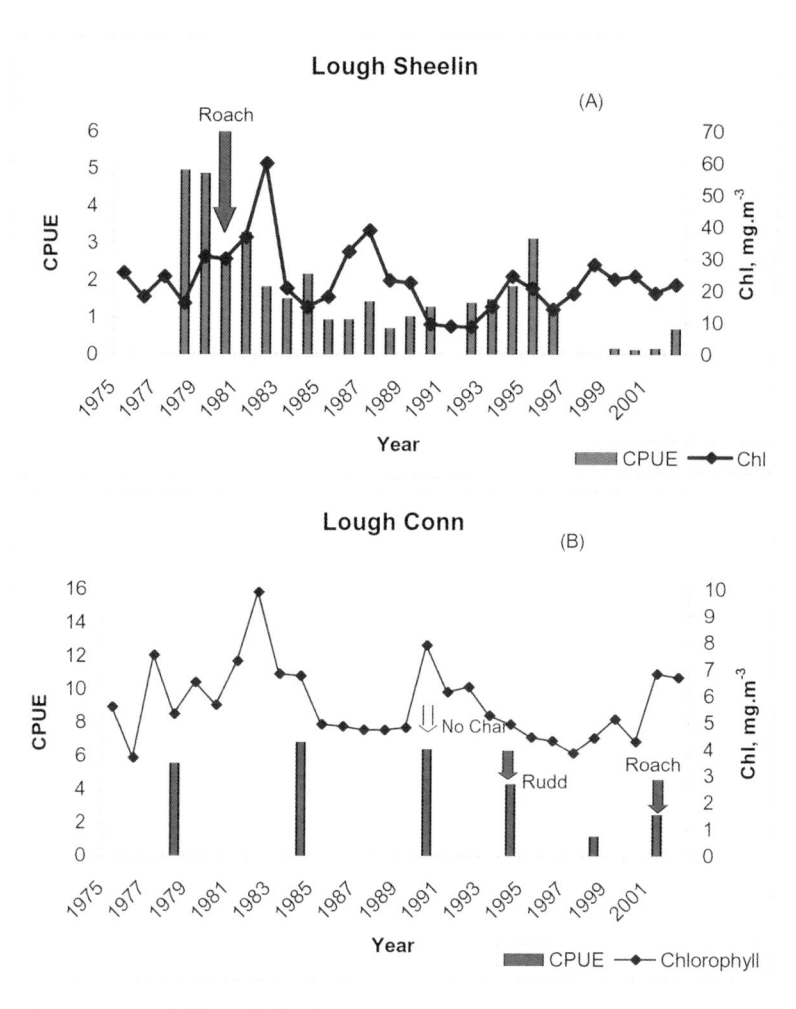

Figure 8: Average annual chlorophyll values (measure of algal biomass) and trout stocks (histograms) in Lough Sheelin (A) and Lough Conn (B) 1975 to 2002. Arrows indicate disappearance of char and appearance of roach and rudd in sampling programme. (CPUE = catch per unit effort).

Open lake monitoring benefits and limitations

The long term monitoring programme of key eutrophication indicators in the open water of selected Irish lakes has significant limitations as a measure of the general ecological "well being" of lakes which are moderately or mildly enriched. The open lake monitoring programme has not been sufficiently sensitive to reflect the significant changes in the fish stocks in Conn. This programme shows that total phosphorus (TP), algal biomass (chlorophyll) and water clarity cannot be used as surrogates to monitor the status of the fish communities in such waters. The programme has proved valuable, however, in that it has successfully demonstrated a clear relationship between TP and chlorophyll in Lough Sheelin, $R^2 = 0.79$ (Champ, 1993) and Lough Ennell, $R^2 = 0.835$ (Foy, *et al.* 1996). It was, therefore, possible to develop phosphorus standards (<20 mg m^{-3} P for mesotrophic lakes) for the protection of Irish alkaline waters (Department of the Environment and Local Government, 1998) giving partial affect to the European Directive on Dangerous Substances (European Council, 1976). It also provides a verifiable record of key lake water quality indicators dating back to the 1970s. Data sets of this type are now especially valuable following the enactment of the Water Framework Directive (2000/60/EEC) European Council (2000). This Directive requires details of a range of parameters for waterbodies of high ecological status (reference conditions) against which to assess current ecological status. In this context also the records provide robust data sets which can be further analysed as more knowledge pertaining to the ecology of these lakes unfolds in future years.

Long-term sea trout rod catch data

Sea trout rod catches are available from the Connemara District of the Western Region since 1974 (Figure 9). Rod catches of about 10,000 fish were recorded from 1974 up until 1986 after which catches fell considerably in 1987 and 1988. Rod catches collapsed to a few hundred in 1989 and 1990 and this collapse was confirmed by returns from fish traps to have been a stock collapse. Post 1990, the introduction of a catch-and-release bye-law and a much reduced fishing effort resulted in very few fish being recorded on rod and line during the early 1990s.

In the context of the Water Framework Directive the long-term monitoring of catches, therefore, was valuable in evaluating historic catch levels (reference conditions), it served to highlight the collapse in stocks in 1989/90 and it will set targets for recovery of stocks in the future.

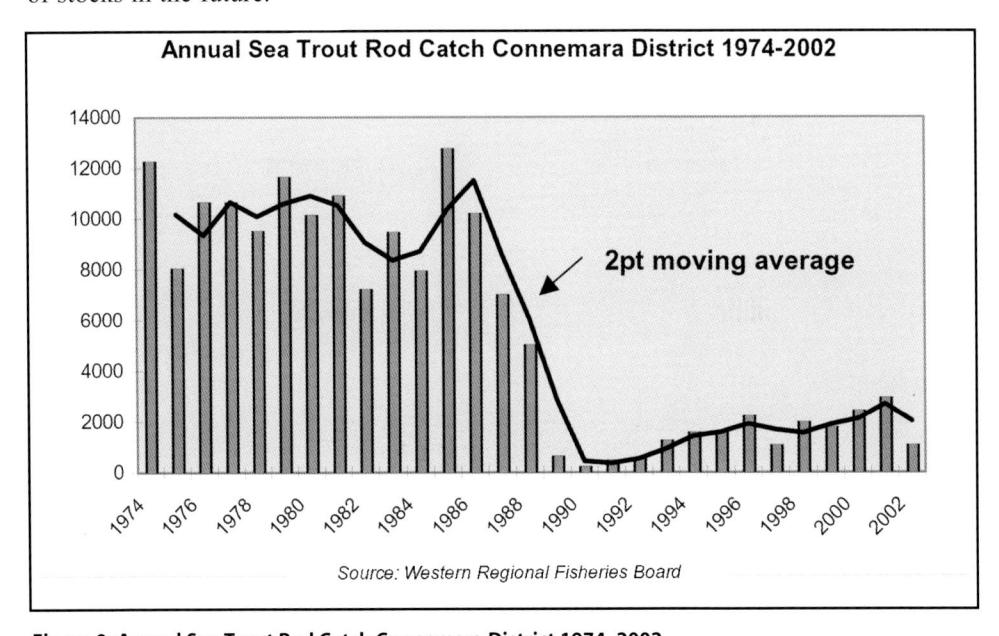

Figure 9: Annual Sea Trout Rod Catch Connemara District 1974–2002

Monitoring of the sea trout stock structure on the Tawnyard subcatchment of the Erriff Catchment, Co. Mayo.

Monitoring of the sea trout stock structure on the Tawnyard subcatchment of the Erriff Catchment has been undertaken since 1985. Length frequency distribution of sea trout kelts[1] are shown (Figure 10). Over the 1985–1988 period, prior to the recorded sea trout collapse in Connemara sea trout fisheries, the population structure of Tawnyard sea trout was typical of a west of Ireland sea trout stock. The population structure was dominated by a finnock[2] peak in the range 27–33 cm and a peak of one-sea winter fish in the range 36–42cm with a representation of older multi-spawners up to 54cm. This population structure collapsed over the 1990–1991 period and thereafter the population was only represented by finnock up until 1993. Over the 1994–1999 period the numbers of sea trout finnock kelts increased and a greater representation of one sea winter and older sea trout became established indicating a partial recovery of the population structure. The collapse in population structure is believed to have been related to mortality of sea trout post-smolts[3] due to sea lice infestation from a marine salmon farm in the estuary. A reduction in sea lice infestation on the estuarine salmon farm from the

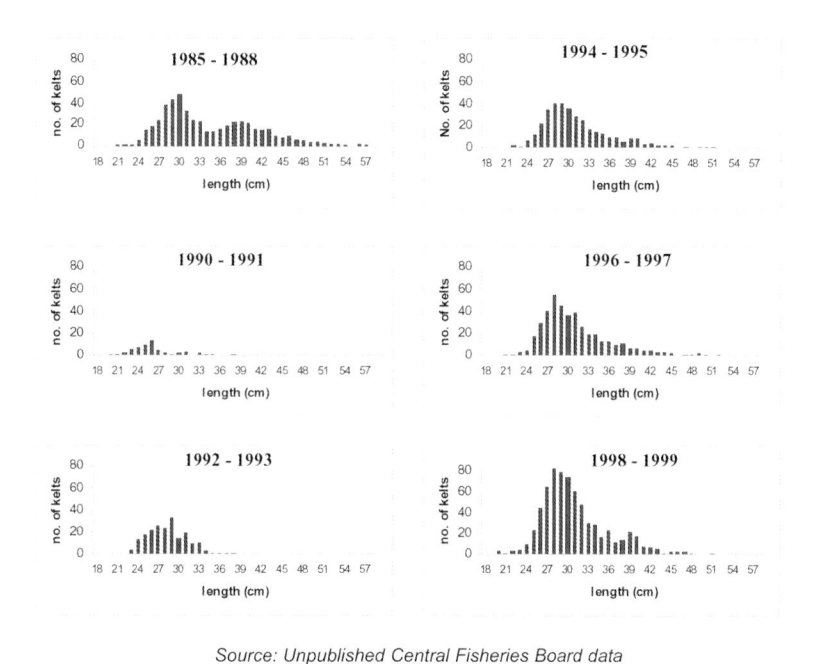

Source: Unpublished Central Fisheries Board data
(1985–1988 data, Whelan & O'Farrell)

Figure 10: Length frequency distribution of sea trout kelts on the Erriff Fishery

---

[1] "kelts" – spent adult fish after spawning
[2] "finnock"– sea trout which return to their natal river within their first 6 months at sea
[3] "post-smolts"– smolts which have recently entered the sea

mid 1990s coincided with the partial recovery of the sea trout population structure on the near-by Tawnyard section of the Erriff catchment.

Long-term monitoring of the sea trout stock allows an assessment to be made of the present status of the population relative to the stock structure in its reference state, prior to the sea trout stock collapse of 1989/1990.

Monitoring of sea lice levels on sea trout around the Irish coast 1992–2001

The relationship between sea lice infestation on sea trout with distance to salmon aquaculture sites for a broad geographic range of Irish rivers was examined over a ten year period, (Gargan *et al.*, 2003). Highest mean levels of total lice and juvenile lice were recorded at sites less than 20 km from farms. The mean total lice infestation decreased at sites between 20 and 30 km from farms and beyond 30 km very low mean total lice levels were recorded. A model was fitted to pooled 10-year data time series for sea lice infestation and distance from marine salmon farms to indicate an overall relationship that could be used to support management actions. Infestations at distances greater than 25 km, never reached over 32 lice per fish and were usually much lower. At distances less than 25 km the full range in infestation occurred. The long-term monitoring programme allowed conclusions to be reached regarding infestation trends and distance to marine salmon aquaculture installations. The results of the monitoring programme led to recommendations being made on the need for effective control of sea lice on marine salmon farms to prevent sea lice infestation in adjacent rivers.

Fish as environmental indicators

Traditionally in Ireland water quality has been assessed in rivers and lakes using various biological (macro-invertebrates and algal biomass) and physio-chemical parameters. No systematic programme is in place for the assessment of fish stocks in rivers or lakes. It is evident from the long term monitoring of trout stocks in Lough Sheelin described above (Figure 8) and the information gained from the limited intermittent studies on Lough Ennell, Corrib and Conn that long term monitoring of fish stocks can provide evidence of environmental change (Champ, 1979; Champ, 2003). The sea trout monitoring programme also provides positive evidence in this regard. The Water Framework Directive now requires Member States to develop and implement a programme of ecological classification which includes fish community composition and age structure of the population. This requirement (to include fish as indicators) demands the introduction of a systematic programme of fish stock assessment.

Marine Sport Fishes

The Central Fisheries Board with the assistance of skippers of marine angling charter boats, have continued a Marine Sport fish Tagging Programme initiated by the Inland Fisheries Trust in 1970. At that time, virtually all fish captured by anglers were killed and taken ashore for weighing. The tagging programme was introduced as a conservation measure and to learn more about the migratory patterns of sea angling species. Sea angling is worth an estimated €30 million to the Irish Economy annually.

Figure 11: Blue Shark recaptures 1970–1998.

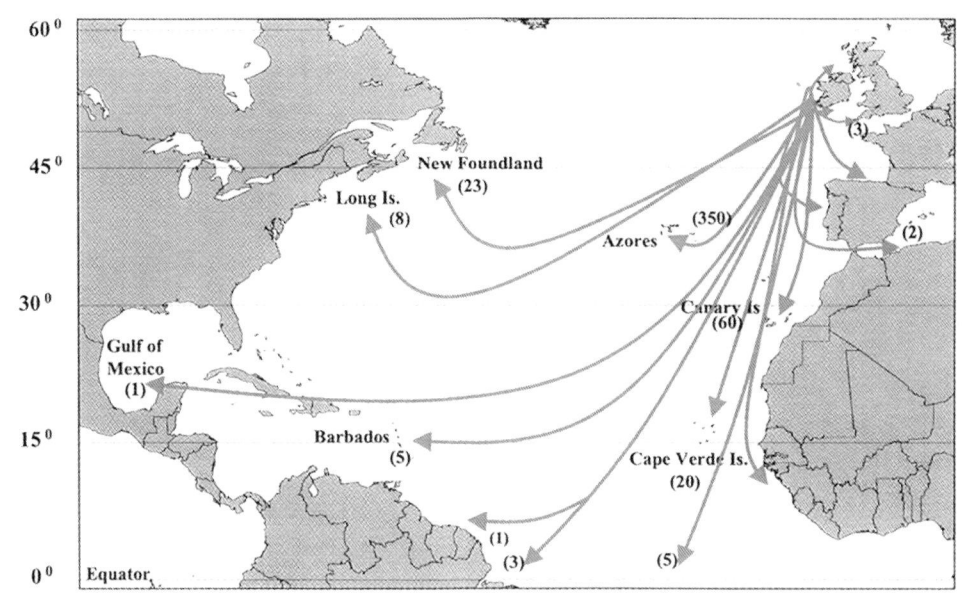

Source: Fitzmaurice & Green, 2000

In excess of 70 charter skippers currently participate in the Tagging Programme. To date approximately 35,000 fish have been tagged and released, blue-shark (17,000) thornback ray (7612), tope (3791) account for the bulk of the tagged fish (Table 2). Recapture details show blue shark tagged off the Irish coast disperse throughout the North Atlantic Ocean (Figure 11). Tope likewise disperse throughout the North East Atlantic from the Faroe Islands to the Azores and Canaries. The rays by comparison appear to remain close to the Irish coast.

The Marine Sport Fish tagging data are currently being analysed as part of a joint collaborative programme between the United States of America and Ireland. The environmental significance of this unique comprehensive data set will emerge in time. The programme has contributed significantly to international understanding of fish stock composition and movements of important commercial and angling species throughout the North Atlantic Ocean.

Conclusions

The long-term monitoring programme of the key water quality indicators of eutrophication shows interesting developments in the large alkaline lakes of the midlands and west. The data for the midland lakes e.g. Sheelin and Ennell show dramatic variation over the thirty year survey period both lakes depicting mesotrophic (moderately enriched) through to hypertrophic (excessively enriched) status at some stage. A strong correlation exists between algal content of the open water of these lakes and total phosphorus concentration. This has contributed to the development of water quality standards for phosphorus (S.I. 258/99). Also as the algae proliferated water

Table 2: Tag and Recapture of Marine Species up to December, 2002

Species	No.s Tagged To Dec. 2002	Numbers Recapt. to Dec. 2002	% Return	Days at Liberty	Dist. Travelled (Miles)	Numbers Tagged in 2002
BLUE SHARK	16996	698	4.10%	2-2399	0-4250	277
PORBEAGLE SHARK	68	7	11.47%	71-3947	89-2300	7
THRESHER SHARK	1	0	0%	-	-	0
TOPE	3791	300	7.9%	28-5538	0-2185	237
MONKFISH	1023	187	18.33%	0-4525	0-720	3
COMMON SKATE	610	93	15.24%	10-3275	0-120	74
LONG NOSED SKATE	3	0	0%	-	1	0
WHITE SKATE	21	1	4.76%	975	3	0
BLONDE RAY	349	20	5.73%	26-1871	0-45	44
UNDULATE RAY	1000	53	5.3%	0-2676	0-60	27
THORNBACK RAY	7612	255	3.34%	0-2190	0-80	503
PAINTED RAY	254	12	4.72%	13-1398	0-20	13
HOMELYN RAY	289	11	3.88%	294-1373	0-72	6
BASS	1805	56	3.10%	0-1373	0-88	0
MULLET	311	5	1.60%	0-275	0-200	0
FLOUNDER	287	30	10.40%	0-738	0-3½	0
STING RAY	28	0	0%	-	-	9
BULL HUSS	26	3	11.53%	103-1499	0-125	0
DABS	43	2	4.65%	378	2	0
SMOOTH HOUND	14	0	0%	-	-	0
PLAICE	5	1	20%	70	0	0
SUN FISH	1	0	0%	-	-	0

TOTAL NUMBER OF FISH TAGGED TO-DATE 34537.

transparency declined (turbidity increased) and areas of lake bed colonised by beneficial plants contracted.

In the highly enriched lakes (Sheelin and Ennell) these changes contributed to substantial reductions in trout stocks. In the moderately enriched waters (Corrib and Conn) evidence of increasing enrichment is less pronounced (localised algal accumulations) but is thought to have caused the disappearance of stocks of char in both these lakes. In Lough Conn, which is marginally more enriched than Corrib, trout stocks have declined significantly while angling success indicates a substantial increase has occurred in trout stocks in Lough Corrib over the past decade. Cyprinid species (roach) introduced in the 1970s and now widespread in Corrib have increased significantly in Conn (as have rudd) since the mid 1990s.

The monitoring programme has shown annual average total phosphorus concentrations less than 20mg.m^{-3} are associated with periods of good ecological conditions (low algal densities) in the midland lakes and values of this order persisted throughout the 30 year study period in the Western Lakes. In the latter waters mean total phosphorus <15 mg.m^{-3} may be necessary to preserve char populations. Furthermore in mesotrophic (moderately enriched) lakes, the key indicators of trophic status at open water locations do not reflect changes in fish communities and are inappropriate for that purpose.

Systematic fish stock assessment is not standard practice in fisheries management currently in Ireland. Historically predator control operations (regular gill netting activities annually in spring) provided some qualitative indication as to fish community structure and density. The development of a trout stock survey method in the late 1970s and its application intermittently to selected lakes provides indisputable evidence of fish community structure and trends in those specific water bodies. Similarly such assessments are limited in rivers. The evidence compiled by the Fisheries Boards of the collapse in sea trout stocks dramatically demonstrates the importance of long-term monitoring and provides invaluable evidence of reference conditions on stock levels in selected western streams.

Systematic surveys are required to quantify fish stocks and confirm fish community composition in rivers and lakes. Such surveys deliver fundamental information for sustainable management and development of the fisheries resource. The absence of factual data of this nature has led to an over reliance on anecdotal evidence to substantiate apparent trends in fish stocks.

The requirements of The Water Framework Directive provide the impetus to establish systematic assessment of fish community composition in rivers, lakes and estuaries. Regular monitoring of representative water bodies will deliver scientific evidence on which to establish future trends and assist with the implementation of measures necessary to ensure legislative compliance and the return of impacted water bodies to good ecological status.

The long-term tagging programme operated by the Central Fisheries Board, with the voluntary participation of charter boat skippers and concerned anglers, provides the only systematic monitoring to date of many important inshore sea anglers fishes. The database which contains records of approximately 35,000 individual fish is the second largest tagging data set in the North Atlantic Ocean. The programme has demonstrated the trans-Atlantic movement of blue sharks. The full environmental significance of the data set is yet to be established and is currently being analysed as part of a collaborative research programme between Ireland and the United states.

Acknowledgements

The authors wish to thank the Central Fisheries Board for permission to present this material. We are especially grateful to Ms. F. O'Connor, Ms. S. McDevitt and Ms. K. Delanty for their assistance with data compilation and to Ms. E. Clarkson for typing the script.

Trevor Champ joined the Fisheries Service in 1968. His primary field of research has been eutrophication induced changes on aquatic ecosystems and in particular the adverse impacts of this phenomenon on salmonid fisheries. He has contributed to the eutrophication debate at Inter-Departmental level; he is Chairman of the Teagasc Environmental Research and Development Advisory Committee. Mr. Champ is responsible for the operation of the Central Fisheries Board laboratory, which specialises in the analysis of nutrients and trace elements in surface waters. He is currently involved in implementation of the Water Framework Directive at Technical Working Group and River Basin District Steering Committee level. He is the co-ordinator of an EPA ERTI Project (2000-MS-4-M1) which is researching the relationships between fish communities and river water quality to develop survey and monitoring protocols for the Water Framework Directive. He served as a National Delegate on Working Group 4 "Impacts of Diffuse Phosphorus on Water Quality": part of EU COST 832 Action on Quantification of the Agricultural Contribution to Eutrophication.

3.7
LONG-TERM MONITORING OF BIRDS IN IRELAND

Oscar J. Merne
National Parks & Wildlife Service, Department of the Environment, Heritage & Local Government, Dublin 2, Ireland[1]

Introduction

Long-term monitoring of birds in Ireland has a rather short history. Up to 1950 most of the emphasis was on establishing the general status and distribution of birds, with some emphasis on the rare or unusual (Thompson, 1851, Ussher & Warren, 1900, Kennedy *et al.*, 1954). By 1950 there was a growing interest in bird migration and bird observatories had been established in Britain and elsewhere to study this phenomenon. The first such observatory established in Ireland was on Great Saltee Island, Co. Wexford, which made systematic daily observations of birds (mainly migrants) during the spring and autumn migration periods, until autumn 1963 when the observatory closed. Another early bird observatory was established on the Copeland Islands, Co. Down, which is still operating today. Later, an observatory was established on Cape Clear Island, Co. Cork, and that, too, is still operating. Bird observatories also operated for a few years on Tory Island and at Malin Head, Co. Donegal.

The monitoring of Pale-bellied Brent Geese (*Branta bernicla hrota*) commenced in the winter of 1960/61 (see below), and other species of wintering wildfowl began to be censused and monitored from the mid-1960s on. Following national waterfowl census

---

[1]Oscar Merne retired from the National Parks & Wildlife Service in January 2004. His home address is 20 Cuala Road, Bray, Co. Wicklow, Ireland.

programmes carried out for a few years in the early 1970s (Hutchinson, 1979) and the mid-1980s (Sheppard, 1993), permanent waterfowl monitoring was put in place with the establishment of the Irish Wetland Bird Survey (I-WeBS) in 1994/95. This survey covers a wide range of wintering waterbirds – divers, grebes, cormorants, herons, swans, geese, ducks, coots, waders, gulls and terns – and the objective is to obtain counts of these species at least three times each winter (early-, mid- and late-) in order to establish total numbers in Ireland, population trends over time, and to identify the most important regular wintering sites for each species. These data are used for conservation decision making, especially for identifying sites that qualify for statutory conservation designation under national and international laws and conventions (e.g. the 1976 and 2000 Wildlife Acts, the European Union Birds Directive (European Council, 79/409/EEC), the Ramsar Convention (1971).

Ireland's breeding seabirds were also increasingly recognised as important in an international context, especially after the first full survey and census of these was completed in Britain and Ireland in 1969 and 1970 (Operation Seafarer, Cramp *et al.*, 1974). A repeat survey and census was carried out in the mid-1980s (Seabird Colony Register, Lloyd *et al.*, 1991), applying improved census methodologies and placing the breeding seabirds of Britain and Ireland in an international context. Another seabird survey and census was carried out in 1999–2002, titled Seabird 2000 (Mitchell *et al.*, 2004). Because of special difficulties in censusing breeding terns, these were covered separately in two all-Ireland tern surveys carried out in 1984 (Whilde, 1985) and 1995 (Hannon *et al.*, 1997).

In addition to the species and groups of species mentioned above, a number of single-species censuses have been targeted at several Irish birds considered to be of particular conservation concern. These include the Peregrine (*Falco peregrinus*) (almost extinct in Ireland by 1970 due to the use of persistent organochlorine agricultural chemicals, but now fully recovered), the Chough (*Pyrrhocorax pyrrhocorax*) (the majority of whose north-west European population occurs on the Irish coast), and the globally endangered Corncrake (*Crex crex)*, once numerous and widespread in Ireland, but reduced by agricultural intensification to little over 100 calling males and confined largely to three areas. The Corncrake is carefully monitored annually, while the other species are censused nationally at ten year intervals.

In more recent years there has been growing concern about apparent large-scale decreases in formerly common and widespread passerine species, and it was quickly realised that little or no reliable data existed from which decreases or changes could be quantified. In order to rectify this deficiency, and to establish a baseline against which future changes could be measured, a Countryside Bird Survey was established by the National Parks & Wildlife Service and the Irish Wildbird Conservancy, using a stratified sample survey methodology which can be easily replicated each year.

For the Sherkin Island Marine Station workshop and conference on long-term monitoring, three examples of long-term monitoring of Irish coastal birds, with which the author has been closely associated, are given below.

Brent Goose Monitoring

In this section the information presented is drawn from Merne *et al.* (1999) and from personal observations and data.

With about 98% of the entire population overwintering here, Ireland has a special importance for the Pale-bellied Brent Goose which breeds on the tundra of the north-eastern Canadian High Arctic. These geese are found around our coasts from late September to late April, or even early May, feeding on *Zostera* spp., green algae (*Enteromorpha* and *Ulva* spp.), salt marsh plants such as *Puccinellia* and *Festuca*, and, in recent times, on agricultural and recreational grasslands close to the coast. From satellite tracking, ringing recoveries and observations it is known that the geese use staging areas in Western Iceland and on the east and west coasts of Greenland on their long and hazardous spring and autumn migrations between their breeding and wintering grounds.

Thanks to the interest, enthusiasm and foresight of the late Major Robert Ruttledge, monitoring of this almost exclusively Irish wintering population of Brent Geese commenced in 1960–61, and has continued almost annually ever since, making it the longest-running annual monitoring scheme for any bird species in Ireland. When Major Ruttledge discontinued the organisation of the monitoring scheme after some years, the task was taken over by Dr. David Cabot, followed by the author of this paper, and finally by the Irish Wetland Bird Survey (I-WeBS) and the Irish Brent Goose Research Group.

The Brent Goose monitoring scheme has three main objectives: (1) an annual census of the entire Irish wintering population, (2) determination of the annual productivity of the population by establishing the percentage of juvenile geese (i.e. birds hatched the previous summer in the Arctic) in the wintering flocks, and (3) identification of the wintering sites of greatest importance for the Brents (e.g. sites of international importance, where 1% or more of the entire Irish wintering population occurs regularly).

To achieve the first objective an early winter census has been carried out at all sites known to support Brents at that time. Traditionally, the great majority of the geese concentrate in the early winter in Strangford Lough, Co. Down, where there are extensive beds of *Zostera*, much sought after by the geese on arrival. To a lesser extent Lough Foyle (Cos. Derry and Donegal) and the Co. Kerry bays (Tralee Bay and Castlemaine Harbour) support significant numbers of Brents in early winter. Elsewhere the other bays and estuaries around the Irish coast between them hold only a few hundred Brents at this time. So, by concentrating observer effort at the main sites mentioned above it is thought that the great majority of the geese can be censused accurately. In recent autumns there has been a tendency for significant numbers of Brents to stay later than usual at staging areas in Iceland, possibly due to climatic amelioration. In order to ensure such birds are included in the population census, arrangements have been made with Icelandic ornithologists to census them at the same time as the early winter census in Ireland.

To achieve the second objective (determination of annual productivity), very large samples of Brent Geese are aged, both during the early winter and mid-winter census (see below). Juvenile Brent Geese are easily identified by the presence of pale tips to the wing covert feathers, which form two or three pale bars on the folded wings. All the samples combined usually add up to 9,000–10,000 Brents, up to 50% or more of the entire population. Numbers of juveniles are expressed as a percentage of the total number of birds aged, from which it is possible to determine how good or bad the breeding season has been, and to relate this to overall numbers and trends over time. Because of severe conditions in the High Arctic breeding grounds there can be total breeding failure in some years. A series of such years can have the effect of reducing overall numbers of geese in the population, as adult birds lost to the population are not replaced by new recruits. However, one or more "good" years, with up to 47% juveniles, can bring about a recovery, or even an increase above the previous population maximum.

The third objective (identification of main sites throughout Ireland) is achieved by a mid-winter census, usually carried out in mid-January, by which time the big early winter concentrations have dispersed following depletion of the *Zostera* food resource. Brents move down the east coast from Strangford Lough to traditional sites in Cos. Down, Louth, Dublin, Wicklow, Wexford and Waterford, while on the west coast birds are found in mid-winter at a number of traditional sites in Cos. Donegal, Sligo, Mayo, Galway, Clare and Kerry. In all, about 27 sites on the Irish coast are internationally important for Brent Geese, based on their regularly supporting 1% or more of the total population. On this basis the sites have been designated as Special Protection Areas under the European Union Birds Directive, in order to safeguard the sites for future generations of Brents.

This monitoring project has demonstrated changes in the Irish wintering Brent Goose population over more than 40 years. Prior to 1960–61, when the project started, it was thought the population may have been as low as 5,000–7,000 birds, possibly due to a combination of shooting pressure and changes in food resources (e.g. loss of *Zostera* due to a wasting disease). By 1960–61 the species was protected from shooting and numbers had increased to 12,000. Between then and now they have fluctuated between 7,350 and 24,700, broadly in line with levels of annual recruitment. There has been an overall upward trend and in recent years the population has averaged about 20,000 birds. The data gathered since 1960–61 will provide an invaluable historical background for an international management plan for this Brent Goose population which is being drafted at present.

The Rockabill Tern Colony

In this section the information presented is drawn from Casey *et al.*, (1995), Patrick *et al.*, (2002) and from personal observations and data.

Rockabill is a tiny (0.9 ha) granitic lighthouse island lying *ca.* 7 km off the coast of north Co. Dublin. Terns are known to have nested there since at least the mid-19th century, although the colony was severely affected by large-scale egg-collecting and trapping/shooting of adult birds for the millinery trade.

During the 20th century, up to 1989 when the lighthouse was automated, the keepers kept a benign eye on the breeding terns. Because terns are very sensitive to human and other disturbance, and to predation, it was considered necessary to warden the colony after the departure of the keepers, particularly as the island supported one of the largest remaining colonies of the rare Roseate Tern (*Sterna dougallii*). The National Parks &

Wildlife Service and BirdWatch Ireland, supported by the Royal Society for the Protection of Birds, established a tern conservation project in summer 1989, which has continued annually since. Wardens are placed on the island from early May, just as the terns are arriving back from their African wintering areas, to early August, when the last chicks fledge. Their duties are not only to protect the colony from disturbance and predation, but also to manage and improve the nesting habitat, census the nests of the different tern species, record clutch size and measure breeding success. Most tern chicks are ringed each year, and feeding studies are also carried out.

Five tern species breed in Ireland each summer, three of them on Rockabill. Besides the Roseate Terns, there are colonies of Common Terns (*Sterna hirundo*) and Arctic Terns (*S. paradisaea*). These two species are fairly widespread in Ireland, but the Roseate Tern is now confined to Rockabill and one other site in south-east Ireland. All the tern species are listed in Annex I of the European Union Birds Directive (European Council, 79/409/EEC), being species of conservation concern in Europe, and the Roseate Tern is also on the Red List of Birds of Conservation Concern in Ireland. Because of the importance of the tern colony on Rockabill the site has been designated a Special Protection Area under the EU Birds Directive, and a Refuge for Fauna under national legislation.

Roseate Terns appear to have become extinct as a breeding species in Ireland by the end of the 19th century, but they began to re-establish here since 1913. Numbers built up in Ireland to a peak of over 2,000 pairs during the 1960s, but by the mid-1970s the population was in serious decline, apparently due to the loss of two major nesting colonies (in Carlingford Lough and Wexford Harbour) to natural erosion processes, and also because of elevated mortality of young Roseate Terns wintering in West Africa, which were caught and killed in large numbers by local people. However, the then relatively small colony on Rockabill, which held 60 pairs in 1969–70, managed to increase to over 300 pairs by 1988. By then it was the largest colony in north-west Europe, containing a little over 50% of the entire breeding population.

In the fifteen years since 1989 the Roseate Tern colony on Rockabill has increased steadily to the current level of nearly 650 pairs, and it is believed this is due to the very high levels of breeding success (up to 1.7 young per pair, in a species which lays one or two eggs) achieved by the conservation project. Harmful disturbance has been virtually eliminated, avian predation has been greatly reduced, and eggs and chicks have been protected from the elements by careful habitat management and the provision of up to 400 nesting boxes which are used as both nesting sites and chick shelters by a large proportion of the Roseate Terns. Catching and killing of Roseate Terns in West Africa has been reduced considerably by a local conservation and education programme. Rockabill now supports over 70% of the entire north-west European breeding population of Roseate Terns and is now beginning to "export" much-needed birds to struggling colonies in north-east England.

Although Common Terns are much more widespread in Ireland, the total population is less than 3,000 pairs. The colony on Rockabill numbered less than 70 pairs in 1969–70, but had increased to *ca.* 170 pairs by the mid-1980s. While the tern conservation project on Rockabill was targeted primarily at the Roseate Terns, the conservation and management measures put in place have greatly benefited Common Terns as well. Numbers have steadily increased over the last sixteen years to the current level of *ca.* 800 pairs, making Rockabill the largest Common Tern colony in Ireland. This species does not nest in the boxes provided for the Roseate Terns, but these are used as shelters by Common Tern chicks, resulting in high productivity in most years – up to 2.35 young fledged per pair, in a species which normally lays two or three eggs.

While there are historical records of Arctic Terns nesting on Rockabill, the species was absent during the 1970s and 1980s. In 1992 two pairs nested and since then numbers have built to to *ca.* 150 pairs in 2004. Most of the Arctic Terns nest in rather exposed positions on The Bill, a rock adjacent to the main island. Because of this exposure to the elements and to some predation by Herring and Great Black-backed Gulls (*Larus argentatus* and *L. marinus*) the productivity is usually quite low. Therefore the increase is likely to be due to immigration rather than being self-generated.

Incidentally, two other seabird species nest on Rockabill and have benefited from the tern conservation project. A small colony of *ca.* 35 pairs of Kittiwakes (*Rissa tridactyla*) has increased to over 150 pairs, while a handful of nesting Black Guillemots (*Cepphus grylle*) has increased to over 35 pairs.

The careful monitoring of the tern colonies on Rockabill over the last sixteen years has charted the very welcome increase in tern numbers and demonstrated the effectiveness of the conservation and management measures which have been applied there.[1]

Seabirds on Great Saltee Island

In this section the information presented is drawn from Lloyd (1981), Roche & Merne (1977), Perry & Warburton (1976), Ruttledge (1963), Mason (1936) and from personal information and data.

Great Saltee is the larger of two islands (the other being Little Saltee) lying *ca.* 7 km off the south Wexford coast and accessible from Kilmore Quay. Since the late 19th century the island has been renowned for its large numbers and variety of breeding seabirds. However, there is very little reliable quantitative information on the breeding colonies until the 1960s. Strangely, the Bird Observatory based on the island throughout the 1950s and up to 1963 concentrated almost exclusively on recording the migratory birds in spring and autumn, and almost ignored the breeding seabirds. A major exception was made for the embryonic Gannet (*Morus bassanus*) colony, which was founded in 1929 but was very slow to increase and expand: between 1950 and 1960 numbers grew from four pairs to 60 pairs.

The other seabird species nesting on Great Saltee are Fulmar (*Fulmarus glacialis*), Manx Shearwater (*Puffinus puffinus*), Shag (*Phalacrocorax aristotelis*), Lesser Black-backed Gull (*Larus fuscus*), Herring Gull (*L. argentatus*), Great Black-backed Gull (*L. marinus*), Kittiwake (*Rissa tridactyla*), Guillemot (*Uria aalge*), Razorbill (*Alca torda*) and Puffin (*Fratercula arctica*). In addition, Cormorants (*Phalacrocorax carbo*) nested

---

[1] **Editor's Note:** Subsequent to the workshop/conference, Oscar Merne reported the following data for 2004. At Rockabill, the number of boxes is now over 500. Numbers of Roseate Terns have increased to 675 pairs, Common Terns to 1,050 pairs, Arctic Terns to 200 pairs and Black Guillemots to 45 pairs.

on Great Saltee up to the 1940s, and several pairs nested in one season in the 1990s. This species now has large colonies on Little Saltee and the inshore Keeragh Islands.

Ornithologists began to ring small numbers of these seabirds in the 1950s, and in the mid-1960s seabird ringing became more systematic and large-scale. At the same time interest in censusing the seabirds on the island began to increase; in 1969 and 1970 a full census of all species was carried out for Operation Seafarer, the first full census of breeding seabirds in Britain and Ireland. This established a good baseline against which to measure population changes and trends over the following decades. Since then the various breeding species on Great Saltee have been censused at regular intervals and a number of the more significant changes and trends recorded are given here, with some discussion on reasons for these.

a) Gannet

As mentioned above, the Gannet colony on Great Saltee grew very slowly from its foundation in 1929 to 60 pairs in 1960. Over the last 43 years the increase accelerated and the colony now has over 2,100 breeding pairs, with annual increases in recent years of up to 9%. This is a much greater rate of increase than the colony itself can generate by recruitment of its own immature birds surviving to breeding age. Therefore the increase is likely to be due largely to immigration from other colonies – perhaps ones such as Little Skellig (Co. Kerry) which have run out of nesting space. The immigration theory is supported by the fact that a large proportion of first time nesters on Great Saltee are unringed: the majority of Gannet chicks produced on the island each year have been ringed.

b) Shag

The population was estimated to be 500 pairs in the 1940s, but it is not known how reliable this count is. Between 1966 and 1977 numbers varied between 130 and 270 pairs. More comprehensive censuses carried out in 1978, 1979 and 1980 produced 250–270, 273 and 434 pairs respectively, demonstrating considerable variation between years. However, numbers remained near the highest level through the 1980s but then were suddenly reduced to 180 pairs as a result of high adult mortality from starvation due to a long series of spring storms making it difficult for the Shags to obtain enough food. Over the last decade there has been a very slow recovery to 230–240 pairs as recruitment of young to the breeding population has been a little greater than adult mortality.

c) Herring Gull

There is some uncertainty about the accuracy of the Herring Gull censuses made prior to 1978, which suggested that up to 5,000 pairs were nesting on Great Saltee. In 1978, 1979 and 1980 the colony was carefully censused, using tested methodology, and produced totals of 3,000, 2,750 and 2,600 pairs respectively. Since then censuses have been carried out at two to three year intervals and these have showed a dramatic decline in numbers, to only 23 pairs by 2003. It is thought that most of this decline is attributable to heavy mortality of breeding adults, and the consequent loss of their eggs and chicks, due to annual outbreaks of avian botulism since the mid-1970s. This massive decline on

Great Saltee fits in with a national decline in the species of 90% between the Operation Seafarer survey of 1969–70 and the Seabird 2000 survey of 1999–2002. The species, once thought to be "too abundant", now qualifies for inclusion in the Red List of Birds of Conservation Concern in Ireland, having declined by 50% in 25 years.

d) Guillemot

Early counts of Guillemots are difficult to interpret, at least partly because the counters did not distinguish between "pairs" and "individual" birds. The first estimate made in the early 1960s indicated 1,500 "pairs" breeding on the island, which seems much too low a figure. Between 1964 and 1971 a series of counts of individual birds (the now generally accepted count unit for this species) gave fluctuating totals ranging between a peak of 13,300 in 1967 and a low of 4,800 in 1970, following a large scale "wreck" of Guillemots in the Irish Sea in autumn 1969. There followed a noticeable increase and recovery, and in 1978, 1979 and 1980 the population estimates were of 11,000 to 13,800 birds. Since then censuses have been carried out at two to three year intervals and have shown a steady increase in numbers, to 20,250 birds in 2003.

Large numbers of Guillemot chicks (usually between 1,800 and 2,800 per annum) have been ringed on Great Saltee since the early 1970s, and recoveries of some of these have provided much useful information on survival, mortality (and its causes), dispersal and migration, and the impacts of major oil spills on the Great Saltee Guillemot population. In spite of high mortality of Guillemots from Great Saltee in the Bay of Biscay following the sinking of the Erika in December 1999, and off north-west Spain following the sinking of the Prestige in winter 2002/03, numbers at the colony in subsequent breeding seasons have shown a continued increase. An analysis of the ring recovery data shows that most of the birds affected were ones in their first winter. Survival of Guillemots to breeding age is generally very low (*ca.*12%), so most of the young birds killed by these two oil spills would probably have died from other causes instead. By contrast, many of the Guillemots killed in the autumn 1969 Irish Sea "wreck" were adults, and this resulted in a large decrease in the breeding population on Great Saltee for several years after the event.

__Oscar Merne__ retired from the National Parks & Wildlife Service in January 2004 (Ed.). He worked as an ornithologist for this Service and its predecessors since 1968. He was responsible for establishing the Wexford Wildfowl Reserve on the North Slob in Wexford, one of Ireland's most important wintering areas for waterfowl — especially the Greenland White-fronted Goose. After ten years he transferred to the NPWS Research Branch headquarters where he was given national and international responsibilities for bird research and conservation. One of his major achievements was the establishment of a network of 110 Special Protection Areas for birds, under the European Union's Birds Directive. His main research interests have been the status, distribution and ecology of breeding seabirds and migratory waterfowl. He has 200 publications on these and other topics.

3.8
OTTERS AND FISH FARMING
– A GOOD NEWS STORY –

Jane Twelves
Salar Smokehouse Ltd,
Lochcarnan,
Isle of South Uist,
Outer Hebrides, H58 5PD,
Scotland, UK

FISH farming often seems to attract bad publicity from some sections of the media, so I am very pleased to have this opportunity of explaining to you that otters and fish farming is actually a good news story.

Otters are the top carnivores in the food chains of both freshwater and coastal marine ecosystems and as such their presence or absence is regarded as an indicator of the state of aquatic ecosystems. The recent return of otters to rivers so polluted for years that no otter has been able to live there has been hailed as a sign that these rivers have at last recovered. It follows that places where there are high densities of otters have a healthy ecosystem and a food chain capable of supporting a high number of these top carnivores.

My involvement with otters and fish farming began over 29 years ago. Most of my work has been done in South Uist, an island in the Outer Hebrides off the north west coast of Scotland. It has been more the observations of a naturalist over the years than a long term scientific study.

I came to live in South Uist in April, 1974 when my husband, Eric, was given the job of setting up a salmon farm in Loch Sheilavaig, a sea loch on the east coast of the island, for Booker McConnell, a multi-national company. 1974 was in the pioneering era in salmon farming and this was one of the first farms to be set up in Scotland and the first to be set up in the Outer Hebrides.

Although I didn't work at the salmon farm, I was very interested in what was happening and sometimes went down to Loch Sheilavaig. It wasn't long before I saw otters there – but this put me in a real dilemma. I was concerned that, on the one hand, here was the embryonic industry of salmon farming just beginning which, if it proved to be successful, would give a very much needed boost to the local economy, but on the other hand here also were rare animals, otters.

Otters had been widespread and reasonably common throughout the British Isles until the late fifties and early sixties but the sudden and widespread use of persistent pesticides in agriculture decimated their numbers. However small pockets had survived in areas where pesticides had not been used and where the countryside had remained much as it had been for many years. The archipelago of the Outer Hebrides was one such area.

I hoped that there would not be a conflict between otters and salmon farming. I hoped that the otters would not be adversely affected by salmon farming, but equally I hoped that the industry would develop successfully. I decided to monitor the situation, but first I had to learn about otters.

Two men who worked at the fish farm used to trap otters, at one time it was legal to do this and the £5 earned from an otter skin was much needed money. These men knew a lot about otters and they taught me how to recognise the signs of the animals. They showed me holts (the underground burrows where otters live), the runs that otters invariably take when they travel overland and they told me to look for 'green stones' which marked the otters 'toilets' as they called them. Otters are territorial animals and they leave their droppings, or spraints, on these prominent spraining sites.

To understand more about the otters in South Uist I decided to do transects going from west to east across the island, mapping the signs of otters to see if there is a pattern to their distribution and to see if they use the different habitats in any particular way.

Otters are piscivores (fish-eating animals) and it seemed a reasonable assumption that they will be found wherever there is food for them.

The geomorphology of South Uist is fascinating. The island is orientated north-south and is approximately 33 km long and 14 km wide, but from west to east across the island there are dramatic changes in land form. The west coast is an almost unbroken stretch of sandy beach, interrupted only by the occasional rocky headland and it takes the full impact of the pounding of the ocean swell built up over the thousands of miles of the Atlantic Ocean. The white sand is composed of the ground-up remains of sea shells and is highly calcareous. Westerly gales have, over centuries, blown this sand inland to form sand dunes and the 'machair', a strip of land about 1 km wide. (Machair is a Gaelic word which, roughly translated into English, means sea meadowland.) To the east is the 'black land', a rib of land about 1 km wide where the calcareous sand has overblown the underlying acid peat to form a soil of neutral pH. To the east again are the wild and rugged acid peat moorlands and hills of South Uist. The east coast is bisected by five fjordic sea lochs, the shoreline here is rocky and, in all but the most exposed areas, is clothed in luxuriant growths of brown seaweeds. There are numerous small offshore islands and skerries. The whole island of South Uist is pitted with literally hundreds of freshwater lochs with pHs ranging from high in the machair lochs to low in the peaty moorland lochs, and there are brackish lochs are the heads of the sea lochs.

On the west coast beaches I found virtually no signs of otters, although occasionally I have seen a set of tracks in the sand.

Inland, among the freshwater and brackish water lochs, I found an amazing and complex network of otter runs, all marked by spraintinig sites, linking all the low-lying lochs and leading to the rocky seashores. I discovered that there is a predictable pattern to this network – otters almost invariably take the shortest route between adjacent lochs or else follow the stream flowing from one loch to another.

It was on the rocky east coast of the island, and particularly in the sheltered east coast sea lochs, where I found most of the holts and other signs of otters – but it was also the sheltered east coast sea lochs in which the salmon farmers were most interested.

I put stretches of wet peat on otter runs and checked them daily for several weeks at different times of year for footprints. In this way I discovered that the extensive freshwater habitat is used mainly during the summer months and usually by single adult males and sub adult animals. The fish bones in the spraints showed that they feed exclusively on eels (*Anguilla anguilla*) and when the eels hibernate in the loch beds in the winter and are difficult to catch, these otters return to the rocky coast to feed and here they have a much more varied diet.

Most of the female otters, with their dependent cubs, live throughout the year on the rocky seashore fishing in the sea, because it is here that there is the best year-round food supply.

The diet of otters living on the rocky coast consists mainly of butterfish (*Pholis gunnellus*) (55% of the diet throughout the year), supplemented with species of wrasse, pipefish, scorpion fish, small flatfish, gadoids and sea sticklebacks (*Spinachia spinachia*).

I soon realised that there is a direct relationship between the standing crop of brown seaweeds, both intertidal and subtidal, and the number of otters. The reason for this, of course, is food supply. Otters forage in the seaweed for their prey, the small fish that they catch they can eat at the surface whereas they take the bigger fish to the nearest reef and shore to eat.

Otters which fish in the sea need fresh water for drinking and rinsing their fur. Otters have no subcutaneous layer of fat to keep them warm; their fur is attached directly to the musculature and they rely on their two layers of fur to keep them dry and warm. The outer hairs are the waterproofing layer and it is the underlying layer of fluffy insulating fur which keeps them warm; it is essential that this layer is kept dry. Salt from the seawater must be rinsed off the outer layer of the fur in order that it maintains its waterproofing qualities. It is noticeable that almost every small pool of freshwater close to the sea is used by otters and marked by spraintinig site(s).

In 1978 I made a map showing the locations of otter holts in the north east of South Uist, an area which included both Loch Sheilavaig, where the salmon farm is located, and Loch Carnan, where Eric and I subsequently set up our own salmon farm.

Loch Sheilavaig is a sheltered loch with a narrow entrance, a complex coastline and many islands, reefs, rocks and underwater undulations. The sea is relatively shallow (but there are no extensive areas of mud at low tide) and seaweed growth is prolific. Loch Carnan is similarly complex and has a similar length of coastline but it has a wide mouth and is more exposed: it has areas of deep water and there are extensive areas of mudflats at low tide. There are more otter holts in Loch Sheilavaig than there are in Loch Carnan.

Later in 1978 a group of fish cages was moored out in North Channel, Loch Sheilavaig, within a few metres of an otter holt on the shore. I thought that if there was going to be a conflict between otters and fish farming it would be here, and there was a chance that this holt might be abandoned. I monitored the holt frequently for a year and less often thereafter. The otters continued to use the holt – and still use it to this day.

From December 1980 to January 1982 I did some radiotelemetry work with Don Jefferies and Tony Mitchell Jones of the Nature Conservancy Council. We caught and radio tagged six otters and two of them lived in Loch Sheilavaig. It was immediately obvious that both of these animals were completely accustomed to the fish farm and to all the activity there, and basically just ignored them. Both animals retained their radio harnesses for approximately three weeks.

An adult male otter lived in a holt on an island near the mouth of the loch. The holt was one kilometre from fish cages. On most days it left the holt to fish and feed around an extensive reef about 300 metres away. One night it was located (but not seen) in the vicinity of the fish cages and on another night it was found two and a half kilometres away in the neighbouring sea loch.

Of particular interest was the sub-adult female which inhabited a holt on an island between the shore base and the group of cages in North Channel, which is approximately 300 metres from the holt. Fish farm boats pass close to the island every day and men working on the cages and at the shore base could be seen and heard from the island. During the three weeks that it wore the radio harness it stayed within a small area close to its holt fishing round the shore of the island and round nearby reefs. The only reaction I saw it make to the activity at the fish farm was one day when it was on a reef near to its holt, it had just finished eating a fish when a fish farm boat came very close. The otter slipped into the water and lay under the seaweed until the boat had just passed, then it came out of the water and carried on as though nothing had happened.

Neither otter predated the fish farm, and I did not see any evidence from spraints or hear from the men working at the fish farm that any of the otters in the loch ate any salmon.

In 1983, nine years after the fish farm in Loch Sheilavaig began, I was satisfied that the industry was not having an adverse effect on the otters and Eric and I set up our own salmon farm in Loch Carnan. Then I started monitoring the holts in Loch Carnan and well as continuing to monitor those in Loch Sheilavaig; all the holts have continued to be used as before.

In 2001, Marine Harvest, who now own the salmon farm in Loch Sheilavaig, asked me if I would do another otter survey of the loch as part of their Environmental Impact Assessment. I carried out the work at the same time of year as I did the survey in 1978, in September.

Although the current generation of otters was not using the habitat in precisely the same way as their ancestors, there were just as many signs of otters in Loch Sheilavaig in September 2001 as there had been in September 1978. All the holts were being used, there was fresh spraint on the sprainting sites, there were footprints in wet peat and crushed vegetation – a lot of evidence of very recent otter activity.

But although the situation regarding the otters was the same, the situation regarding the fish farm had changed; it had expanded considerably over the intervening 23 years. There were now three large groups of cages in the loch, North Channel, Hole Bay and North Bay, and each of these groups of cages has an active holt within a few metres of it. There is also a holt close to the shore base and boat moorings. Otters can, therefore, obviously tolerate having fish farm cages, boat moorings and shore bases very close to their holts.

I then asked the question, 'How much salmon has been harvested from Loch Sheilavaig since the salmon farm was set up?' and it was calculated that in excess of 10,000 (ten thousand) tonnes of salmon had been harvested from the loch between 1974 and 2001. If we assume an average food conversion ratio of 1.3:1 during this time, we can quickly calculate that 13,000 (thirteen thousand) tonnes of fish food had been put into the loch and 10,000 tonnes of salmon taken out.

All this fish farming activity in Loch Sheilavaig continuously over 27 years has not affected the top predator in the marine ecosystem, the otter. The presence of otters in aquatic habitats is taken as an indication that the habitat is in a healthy state. The otters in Loch Sheilavaig, and other sea lochs where there are fish farms, have been the independent long-term monitors of the environment of the lochs and their continued presence means that the fish farms are not having the dire effect on the environment that many media reports would have us believe.

This is not really surprising because salmon demand clean, unpolluted water otherwise they fail to grow and the fish farm business fails. Otters and salmon are each, in their different ways, canaries of the marine environment in which they live.

In conclusion, I am happy to be able to tell you that otters and fish farming is indeed a good news story – and it is important for us to know this because fish farming is now one of the main stays, if not *the* main stay, of the Scottish rural economy.

Jane Twelves moved to South Uist in 1974 and began monitoring otter populations in relation to the developing fish farming industry. For the last 20 years she has worked as a fish farmer and fish smoker.

3.9
Thirteen Years of Monitoring Sea Lice in Farmed Salmonids

David Jackson
Lorraine Copley, Frank Kane,
Oisín Naughton, Suzanne
Kennedy & Pauline O'Donohoe,
Marine Institute,
Galway Technology Park,
Parkmore, Galway,
Ireland

Introduction

Salmonids farmed in Ireland can be divided into the following groups: one year class of rainbow trout (*Oncorhynchus mykiss*) and three year classes of Atlantic salmon (*Salmo salar*). The year classes of salmon include, smolts, one sea-winter salmon and two sea-winter salmon. Sfi's are fish which are transferred to sea in Autumn/Winter of the same year that they are hatched. Their S1 siblings smoltify and are put to sea in early spring, some three to four months later. Salmon which are at sea for a year or longer in April are known as growers/one sea-winter and are treated separately from younger salmon (smolts) and rainbow trout.

Two species of sea lice are found on cultured salmonids in Ireland, *Caligus elongatus* Nordmann, a species of parasite that infests over eighty different types of marine fish, and *Lepeophtheirus salmonis* Krøyer (Caligidae), which infests only salmon and other salmonids. Sea lice are regarded as having the most commercially damaging effect on cultured salmon in the world with major economic losses to the fish farming community resulting per annum (Bristow and Berland, 1991; Jackson and Costello, 1991). They affect salmon in a variety of ways: mainly by reducing fish growth, loss of scales which leaves the fish open to secondary infections (Wootten *et al.*, 1982) and damaging of fish which reduces marketability.

Lepeophtheirus salmonis is regarded as the more serious parasite of the two species and has been found to occur most frequently on farmed salmon (Jackson and Minchin, 1993). Most of the damage caused by these parasites is thought to be mechanical, carried out during the course of attachment and feeding (Kabata, 1974; Brandal *et al.*, 1976; Jones *et al.*, 1990). Inflammation and hyperplasia (enlargement caused by an abnormal increase in the number of cells in an organ or tissue) have been recorded in Atlantic salmon in response to infections with *L. salmonis* (Jones *et al.*, 1990; Jonsdottir *et al.*, 1992; Nolan *et al.*, 2000). Increases in stress hormones caused by sea lice infestations have been suggested to increase the susceptibility of fish to infectious diseases (MacKinnon, 1998). Severe erosion around the head caused by heavy infestations of *L. salmonis* has been recorded previously (Pike, 1989; Berland, 1993). This is thought to occur because of the rich supply of mucus secreted by mucous cell-lined ducts in that region (Nolan *et al.*, 1999). In experimental and field investigations carried out in Norway heavy infestation was found to cause fish mortalities (Finstad *et al.*, 2000).

Lepeophtheirus salmonis has a direct life cycle, meaning it has a single host. There are ten stages in the life cycle of *L. salmonis*, each separated by a moult (Kabata, 1979; Schram, 1993). Moulting involves the shedding of the outer shell or cuticle, to expose a new cuticle underneath. After hatching from the egg (which is extruded from the adult female louse in paired egg-strings) a free-living nauplius stage is dispersed into the water column and survives in the plankton for a short time. This is then followed by a second nauplius stage which eventually moults into a copepodid, which can survive in the plankton for a number of days. This copepodid must locate a salmonid host before the parasite can develop further. Copepodids make initial contact with the host by grasping the surface of the host with their mouthparts and driving the clawed second antennae into the epidermis. Following settlement the copepodid moults into the chalimus phase which comprises four stages, characterised by permanent attachment to the host by a frontal filament (Johannessen, 1978). This frontal filament is lost when the chalimus IV stage develops into the mobile pre-adult male or female. A moult then separates two pre-adult stages after which the fully mature adult develops. The adult female is capable of producing a number of batches of paired egg-strings during her life span, which in turn hatch into the water column giving rise to the next generation. The number of egg-strings that can be produced by an adult female of *L. salmonis* can vary. Ritchie (1993) showed that six pairs of egg-strings were extruded over a period of 50 days at 14°C after one mating. However, Heuch *et al.* (2000) showed that some females, kept at a lower water temperature of 7.2°C, could produce as many as eleven egg-strings, and stated that in the wild this value could be even higher. Various survival times have also been recorded for this species. Earlier studies have given survival times of 75 days at 14°C (Ritchie, 1993) and 191 days at 7°C (Nordhagen, 1997). However, a later study (Nordhagen *et al.*, 2000) indicated that the life span of *L. salmonis* could be up to one year at lower water temperatures.

Caligus elongatus is not as host specific as *L. salmonis* and parasitises a wide range of marine fish (Kabata, 1979). This, combined with the migrating patterns of their hosts, is thought to account for the highly variable levels on farmed salmonids at different times of the year. The developmental stages of *Caligus elongatus* described by Hogans

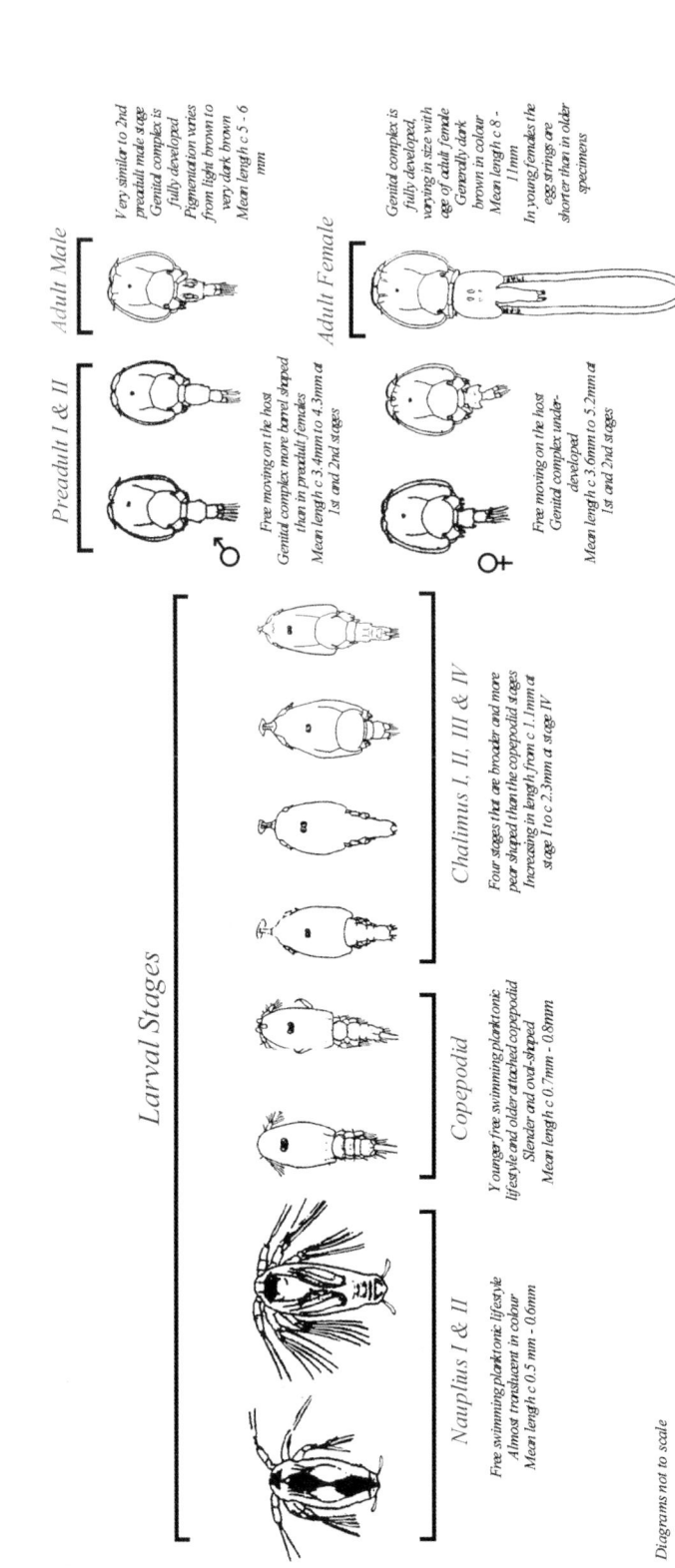

Figure 1: Life cycle stages of *Lepeophtheirus salmonis* (after Schram, 1993)

and Trudeau (1989) included nine stages, each separated by a moult as in *L. salmonis*. These stages included two free-living nauplii, a copepodid, four attached chalimi, a pre-adult and adult. However, studies by Piasecki (1996) contradict these earlier findings and state that there are only eight stages in the life cycle of *C. elongatus*. Piasecki (1996) maintains that previous studies labelled young adult individuals of *C. elongatus* as pre-adult stages. The adult life span for *C. elongatus* has been estimated at 260 days for an adult female in typical winter water temperature ranges of 2.2–12°C. Two sets of egg-strings are believed to be produced following a single mating (Piasecki & MacKinnon, 1995).

There are three licensed treatments for sea lice control in Ireland. Two of these treatments, Calicide and Slice, are in-feed treatments and the third, Excis, is a topical treatment. Calicide contains teflubenzuron which acts as a chiton synthesis inhibitor that interferes with the cuticle formation of the louse. It is only effective against the moulting stages of the life cycle and it has a 7 day withdrawal period. Slice contains emamectin benzoate, which interferes with the peripheral nervous system causing paralysis or death. It is effective against all stages of the life cycle and has no withdrawal period. The topical treatment Excis contains cypermethrin, which again affects the nervous system. It is effective against all stages of the life cycle and has a 24 hour withdrawal period. The alternation between these treatments is an important management strategy to help combat lice resistance to chemotherapeutants.

All finfish farms are obliged to participate in the state run sea lice monitoring and control programme. This involves the inspection and sampling of each year class of fish at all fish farm sites *fourteen* times per annum, twice per month during the Spring period – March, April and May, and monthly for the remainder of the year except December–January. Only one inspection is carried out during this period.

The four purposes of the National Sea Lice-Monitoring and Control Programme are:
- To provide an objective measurement of infestation levels on farms
- To investigate the nature of the infestations
- To provide management information to drive implementation of the control and management strategies
- To facilitate further development and refinement of the control and management strategies.

The sea lice monitoring and control strategy has five principal components:
- Separation of generations
- Annual fallowing of sites
- Early harvest of two sea-winter fish
- Targeted treatment regimes, including synchronous treatments
- Agreed husbandry practices.

Together, these components work to reduce the development of infestations and to ensure the most effective treatment of developing infestations. They minimise lice levels whilst controlling reliance on, and reducing use of, veterinary medicines. The separation of generations and annual fallowing prevent the vertical transmission of infestations from one generation to the next, thus retarding the development of infestations. The early harvest of two sea winter fish removes a potential reservoir of lice infestation and the agreed practices and targeted treatments enhance the efficacy of treatment regimes. One important aspect of targeted treatments is the carrying out of autumn / winter treatments to reduce lice burdens to as close to zero as practicable on all fish, which are to be over-wintered. This is fundamental to achieving zero / near zero egg-bearing lice in spring. The agreed husbandry practices cover a range of related fish health, quality and environmental issues in addition to those specifically related to lice control.

The setting of appropriate treatment triggers is an integral part of implementing a targeted treatment regime. Treatment triggers during the spring period are set close to zero in the range of from 0.3 to 0.5 egg-bearing females per fish and are also informed by the numbers of mobile lice on the fish. Where numbers of mobile lice are high, treatments are triggered even in the absence of egg-bearing females. Outside of the critical spring period, a level of 2.0 egg-bearing lice acts as a trigger for treatments. This is only relaxed where fish are under harvest and with the agreement with the Department of Communications, Marine and Natural Resources (2000) or its agent.

Over the period since the initiation of Single Bay Management (SBM), treatment triggers have been progressively reduced from a starting point of 2.0 per fish during the spring period to the current levels which are the optimal sustainable at present. These trigger levels will be kept under review in the light of advances in lice control strategies. Triggered treatments are underpinned by follow up inspections and, where the Department or its agent considers it to be necessary, by sanctions. Sanctions employed include, peer review under the SBM process, conditional fish movement orders and accelerated harvests.

Methodology

Sampling frequency has been determined with regard to sea lice development rates, critical periods and environmental conditions. During winter when temperatures are lower, (December–February) lice development occurs slowly and a low frequency of inspection will detect changes adequately. During the spring rise in temperatures, lice development accelerates therefore more frequent sampling is carried out.

The sampling frequency is fourteen inspections per year, plus any follow-up inspections required where advice to reduce lice levels has been issued. One lice inspection takes place each month at each site where fish are present, with two inspections taking place each month during the spring period March to May. Only one inspection occurs for December / January. At each inspection two samples are taken for each generation of fish on-site. One is from a standard cage (which is sampled at each inspection) and one from a random cage (which is selected on the day of the inspection). Thirty fish are examined for each sample. These are anaesthetised in a bin, which at the end of the sample is sieved for any detached lice. Each fish is examined individually for all mobile lice. Lice are removed using forceps and placed in 30ml screw top plastic bottles containing 70% alcohol, one bottle per fish. The results presented in this report refer to mean lice numbers per fish. This was obtained by adding the number of lice taken per fish with the number from the bin, and dividing by the number of fish examined.

Monthly reports are compiled of each site for mean numbers of egg-bearing lice and total mobile lice of each species. These reports are circulated to the farms, the Department of the Marine and Natural Resources, the Marine Institute, the Central Fisheries Board, the Regional Fisheries Boards, Save Our Sea Trout, the Western Gamefishing Association and the Irish Salmon Growers' Association. This ensures that detailed information on the levels pertaining on farms is available to all interested parties. These reports are designed to give a clear, unambiguous measure of the infestation level at each site and to act as a basis for management decisions.

There are three regions where salmon farming is carried out, the West (Counties Mayo and Galway), the Northwest (Co. Donegal) and the Southwest (Counties Cork and Kerry). These are geographically separate from each other with distances between regions of *ca.* 160 km from Northwest to West and *ca.* 200 km from West to Southwest. In the year 2002 a total number of 45 sites were inspected around the west coast of Ireland (See Figures 2–5).

Results presented are mean ovigerous sea lice levels (egg-bearing adult female lice) and mean mobile sea lice levels (lice that have developed beyond the attached chalimus stages) for *Lepeophtheirus salmonis* and *Caligus elongatus*. Total mobile levels estimate successful infection, with ovigerous lice levels estimating successful breeding females. The regularity of the monitoring protocol outlined above aims to evaluate the levels of lice on growing fish and to bring them under control if necessary by advising treatment. Effective parasite control is characterised by a drop in lice levels in the subsequent inspection.

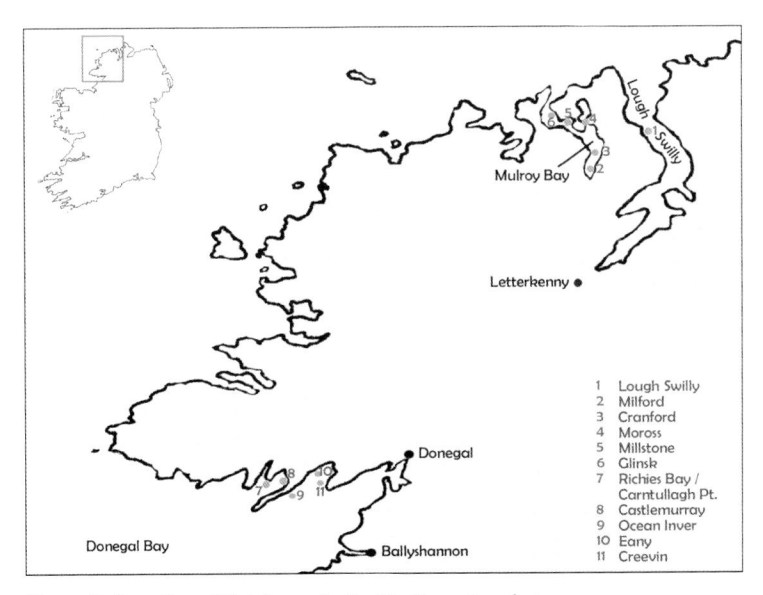

Figure 2: Location of fish farms in the Northwest region.

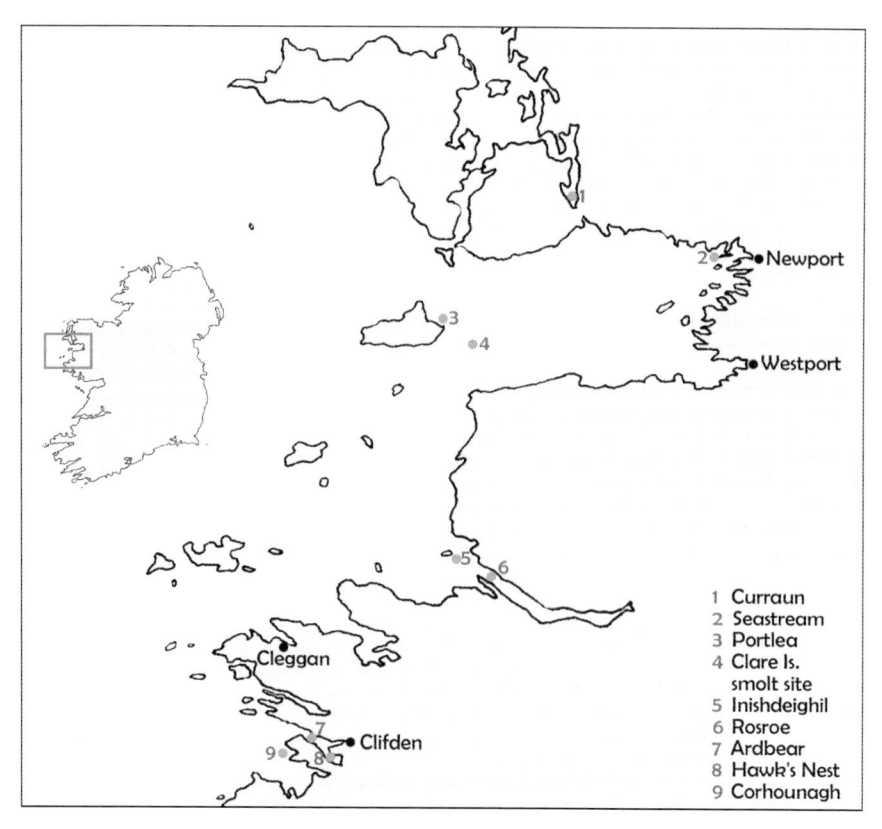

Figure 3: Location of fish farms in the Western region (Clew Bay / Connemara).

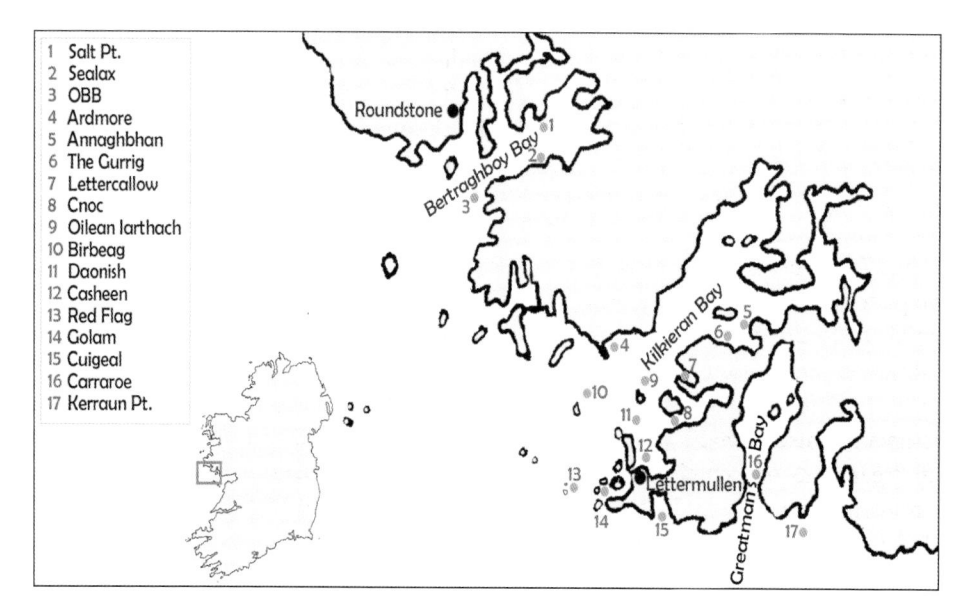

1 Curraun
2 Seastream
3 Portlea
4 Clare Is. smolt site
5 Inishdeighil
6 Rosroe
7 Ardbear
8 Hawk's Nest
9 Corhounagh

Figure 4: Location of fish farms in the Western region (Connemara).

1 Salt Pt.
2 Sealax
3 OBB
4 Ardmore
5 Annaghbhan
6 The Gurrig
7 Lettercallow
8 Cnoc
9 Oilean Iarthach
10 Birbeag
11 Daonish
12 Casheen
13 Red Flag
14 Golam
15 Cuigeal
16 Carraroe
17 Kerraun Pt.

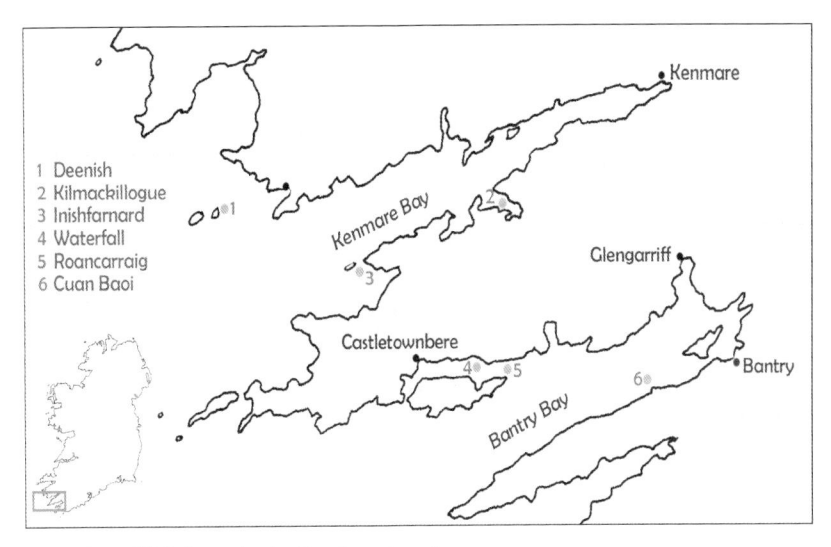

Figure 5: Location of fish farms in the Southwest region.

Results

Each year the National Monitoring Programme carries out between 450 and 500 inspections of fin-fish farm sites. These comprise inspections of smolts in their first year at sea, inspections of grower fish of previous generations and inspections of farmed rainbow trout. The results of each inspection are relayed to the farms within five working days of the inspection and feed into the farms sea lice control plan. If the inspection reveals that levels have reached the treatment trigger levels an immediate treatment is advised. If infestations are below this level the information, together with the information on infestation levels at adjacent sites, is used to plan strategic treatments on a bay-wide basis. Over the years this approach has enabled farmers to progressively improve their control of sea lice (Figure 6). A thirteen-year trend graph of all data for May of each year shows that the mean number of the salmon louse, *Lepeophtheirus salmonis*, on one-sea-winter farmed salmon is going down.

Looking at the ratio of adult female *L. salmonis* to total mobile lice over the years (Figure 7) in three key bays, Donegal, Kilkieran and Kenmare, it is clear that there are two patterns. In one case the numbers of total mobile and ovigerous female lice are very similar and remain so over most of the year. In the second case the numbers of total mobiles are much higher than the numbers of ovigerous females. These patterns reflect the type of lice management regime in place. Where in-feed treatments are the main therapeutant used, total mobile and ovigerous numbers both tend to be similar. Where bath treatments or a combination of bath and in-feed treatments are used there can be a significant difference in the levels of ovigerous females and total mobiles.

The monitoring programme was based on taking two large samples of fish (thirty fish) at each site inspected. One sample was taken from a standard cage and the other from a cage picked at random to ensure that the results from the standard cage were representative. Analysis of the data for the years 1995 to 2002 (Figure 8) using Mann

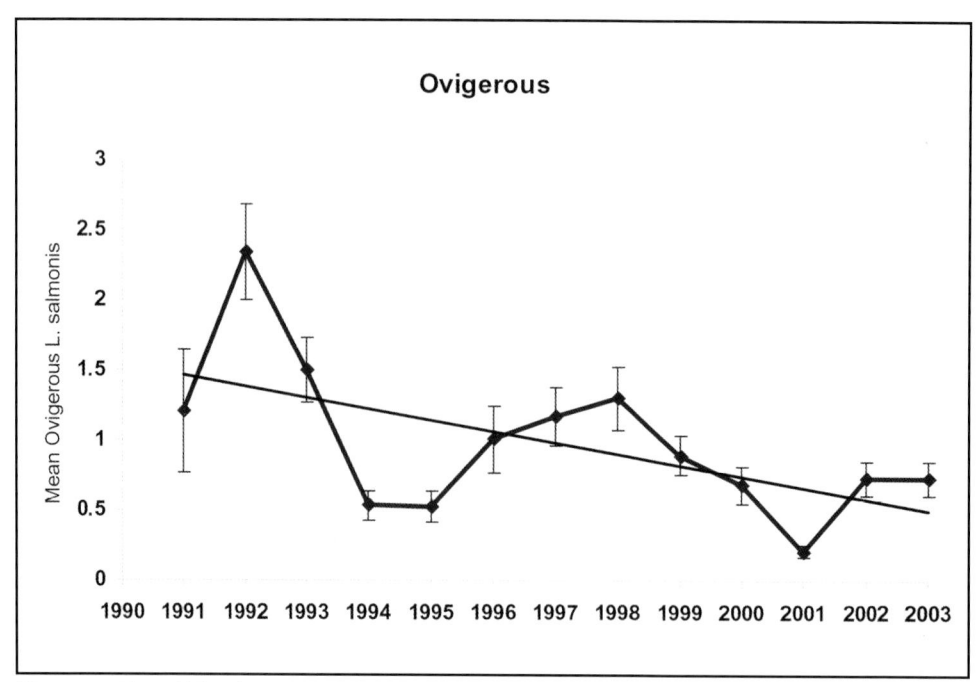

Figure 6 a: Mean (SE) ovigerous *Lepeophtheirus salmonis* on one sea-winter salmon in May of each year with trendline.

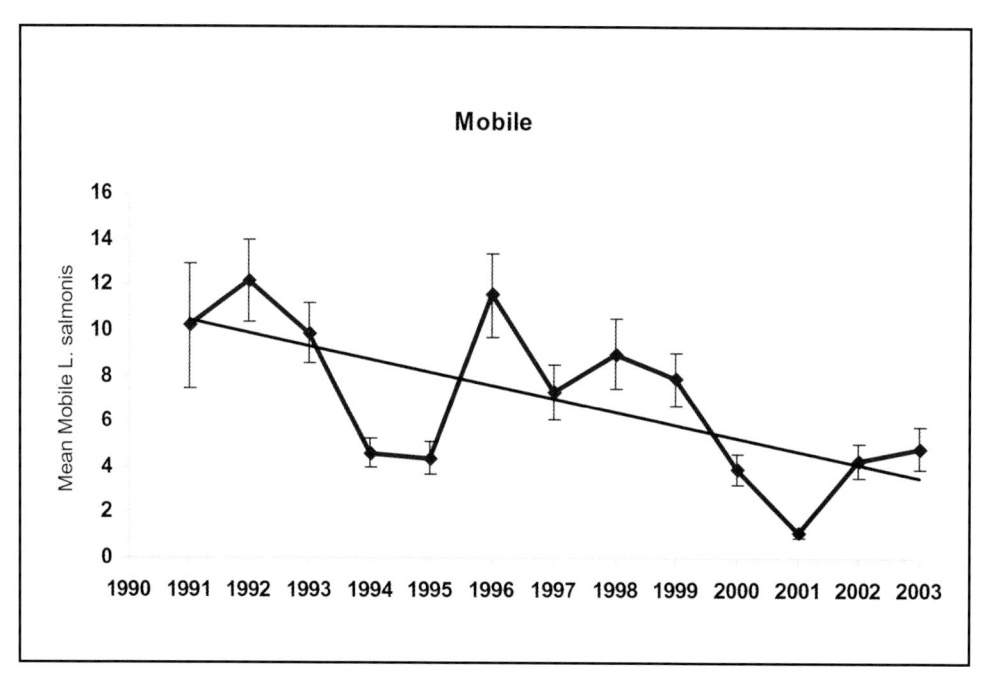

Figure 6 b: Mean (SE) mobile *Lepeophtheirus salmonis* on one sea-winter salmon in May of each year with trendline.

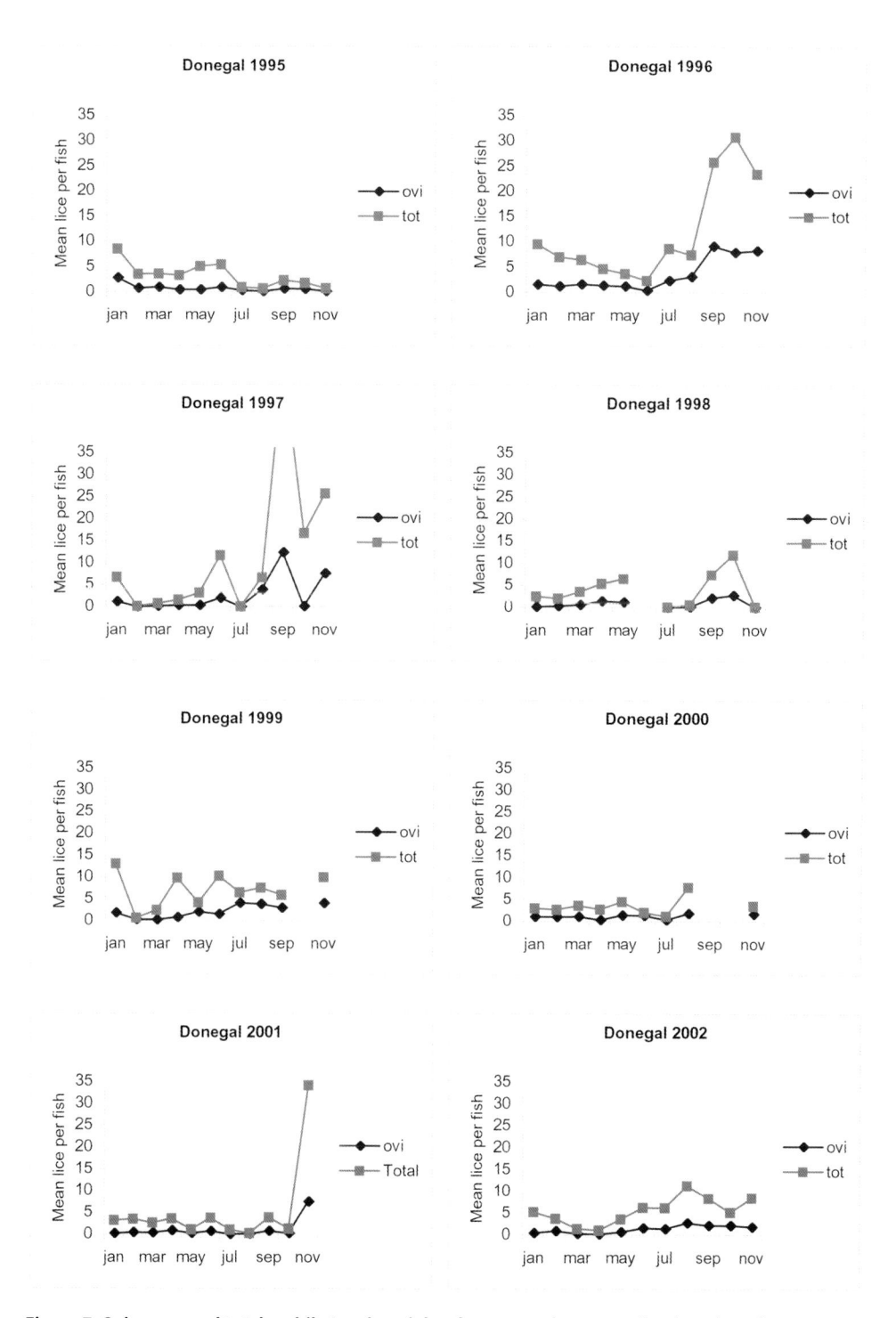

Figure 7: Ovigerous and total mobile *L. salmonis* levels per month per year for three bays from 1995–2002.

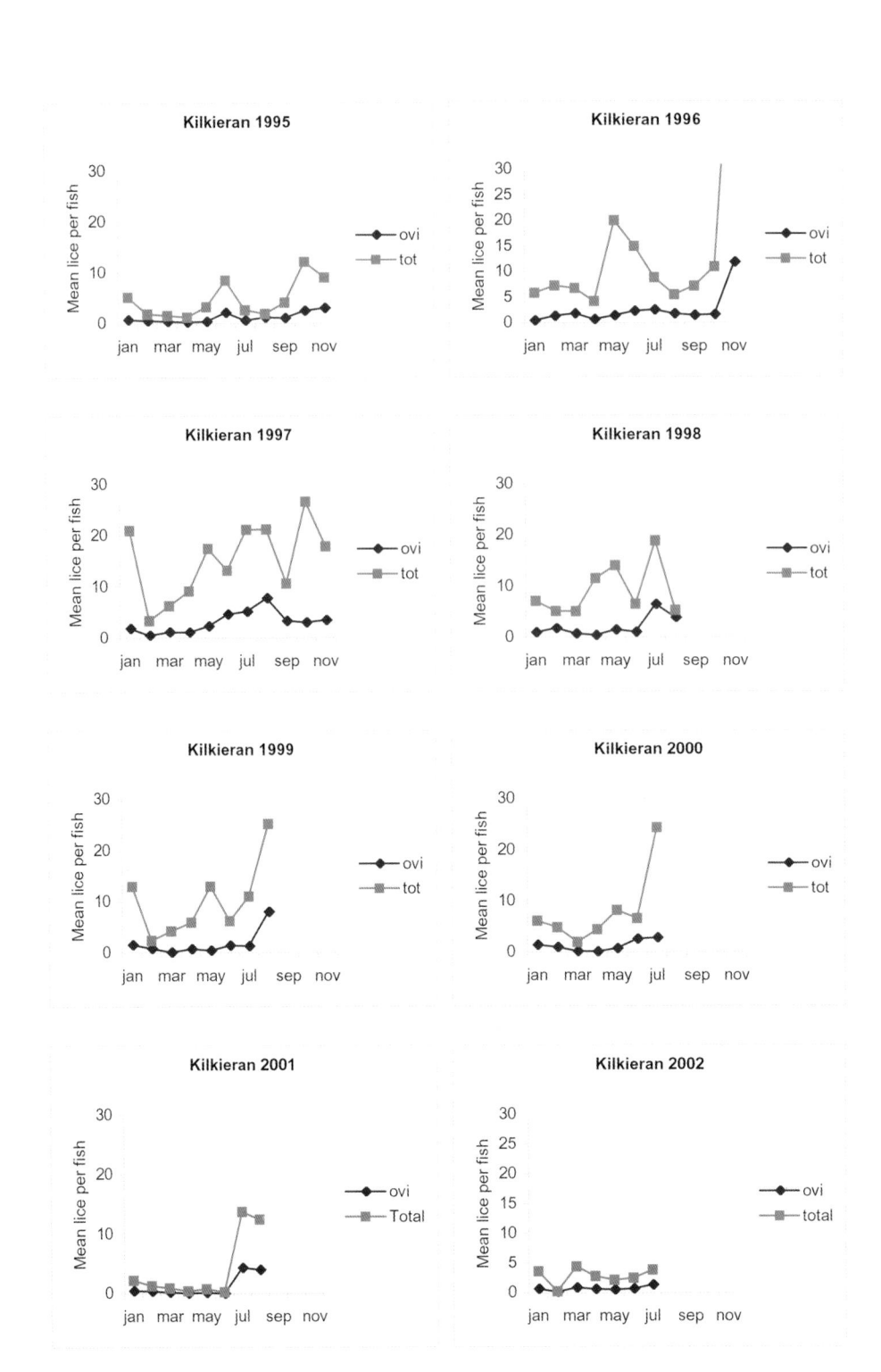

Figure 7 continued

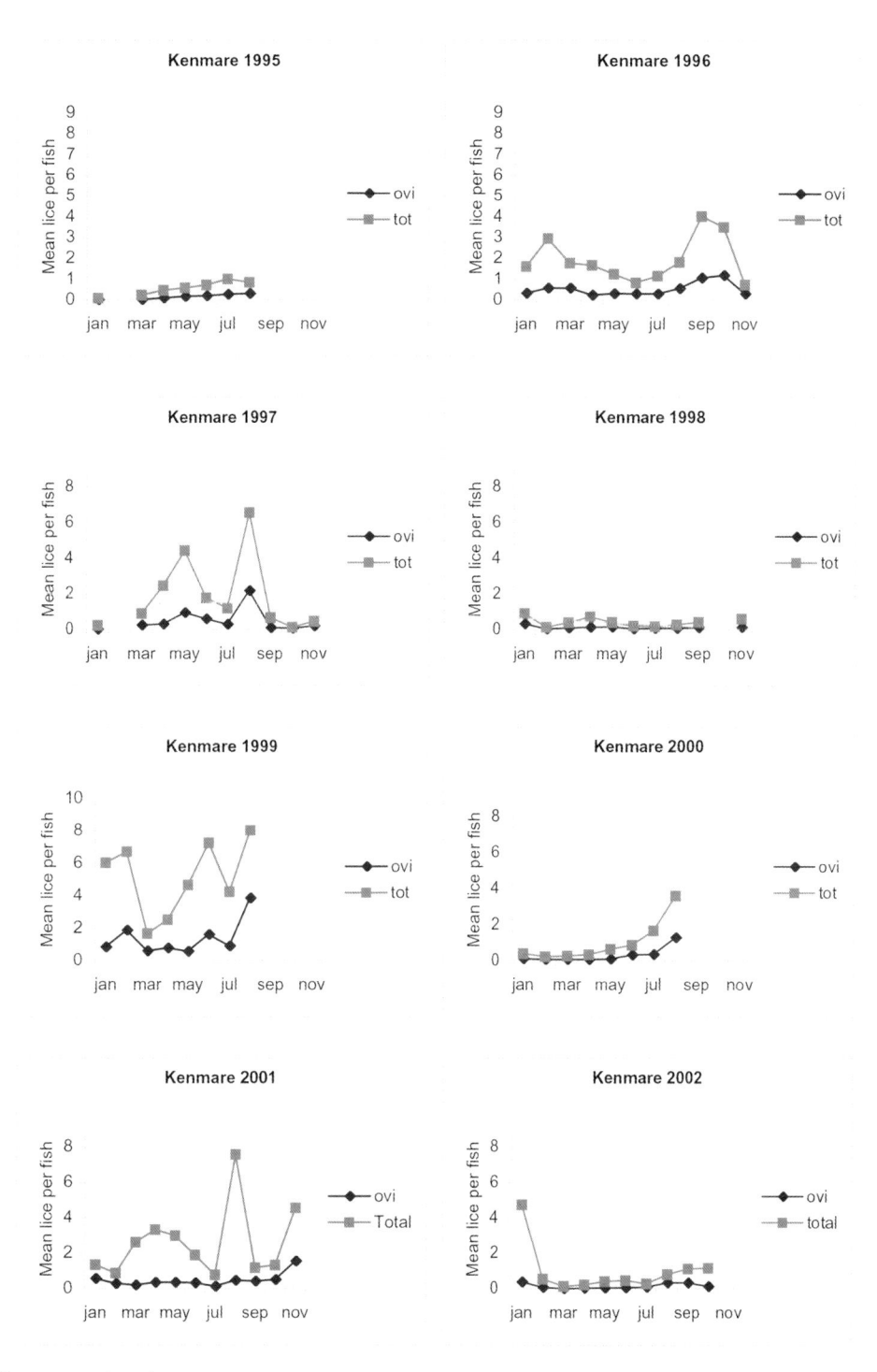

Figure 7 continued

Whitney U tests shows that there is no significant difference between lice levels in standard and random cages.

Discussion

A cornerstone of sea lice management in Ireland is the concept of SBM, which is a locally based, co-operative approach to management of sea lice and disease within a bay. It relies on integrated management rather than treatment to control sea lice.

One such practice is the annual fallowing of production sites. A bay or site is considered fallow when there are no fish stocks present. A minimum of 30 days is required. In studies in Ireland and Scotland, it was shown that a longer period of time was taken for significant infestation to build up on new smolts put to sea following a

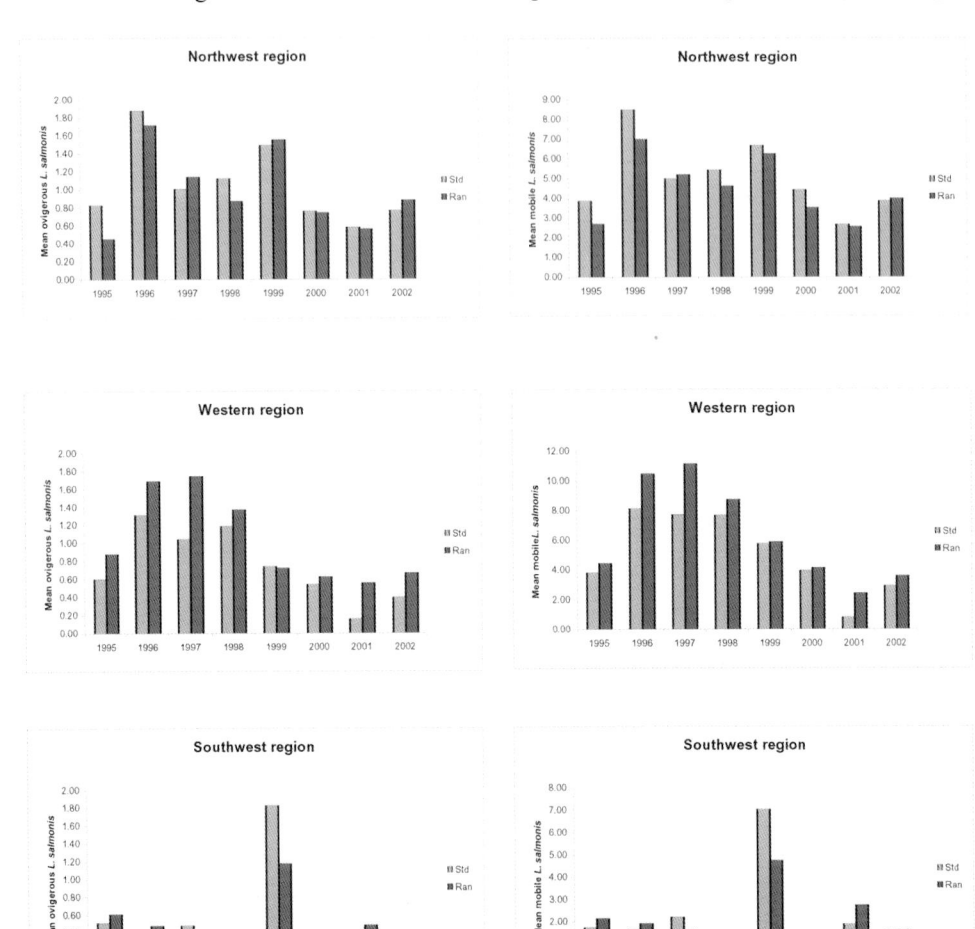

Figure 8: Comparisons between standard and random cages for ovigerous and mobile *L. salmonis* from 1995–2002. No significant differences were recorded between standard and random cages using Mann Whitney U tests.

fallow period, thereby delaying the need for chemical treatment (Jackson & Minchin, 1993; Grant & Treasurer, 1993). Mixed-generation sites also showed rapid infestation on incoming smolts (Grant & Treasurer, 1993). The practice of fallowing and single-generation sites was seen to reduce the lice burden on farmed salmon smolts going into their first winter along the west coast of Ireland as described in the Report of the Sea Trout Working Group, 1994 (Department of the Marine, 1995). Fallowing and separation of generations of stock on site and within bays has long been used as a key SBM tool in Ireland to prevent cross contamination between fish (Report of the Sea Trout Working Group, 1992, 1993 and 1994) (Department of the Marine, 1992, 1993 and 1995, respectively). The early harvesting of two-sea-winter fish is a recommended SBM element and an important co-operative approach is the establishment of agreed codes of practice between farms to minimise sea lice levels. Studies have shown that significant benefits were achieved when a late winter treatment was administered at a number of fish farm sites in Scotland (Wadsworth *et al.*, 1998). These included a reduction of mobile lice infection levels and a reduction in the future need for treatments. The co-ordination of targeted treatments, synchronised to maximise the effect within a bay and prevent cross contamination from untreated fish is an integral part of SBM. Lower sea lice infestations have been attained in Ireland where the implementation and synchronisation of management practices have been employed (Jackson *et al*, 2002). In Scotland synchronous treatments with adjacent farms in appropriate areas was shown to improve lice control (Wadsworth *et al.*, 1998). By using treatments correctly to their optimum effect, through knowledge of the life cycle of the sea lice, longer intervals between treatments was achieved. In addition, SBM establishes a forum for information exchange and involves holding regular meetings with all producers within a bay to discuss issues and plan management strategies.

The management strategies developed for farmed salmon in Ireland over the past decade were developed based on the results of a solid monitoring programme. This monitoring programme supplied the objective information and data, which allowed management protocols to be developed in the first instance. In the intervening years the monitoring has allowed the decision-making process to be fine-tuned to take account of temporal variations in lice infestation patterns. It has also facilitated further refinement of the management protocols themselves.

***David Jackson** is manager of the aquaculture section of the Marine Institute. His expertise lies in integrated management, coastal zone management, the development of new species for aquaculture and sea lice biology and managment. He is responsible for the sea lice monitoring and control, the single bay management and the CLAMS programmes in the Marine Institute and is involved in a number of international projects on aspects of aquaculture management.*

3.10
THE SEDIMENTARY RECORD SHOWS THE NEED FOR LONG-TERM MONITORING OF PHYTOPLANKTON

Barrie Dale
Dept. of Geosciences,
University of Oslo,
PB 1047 Blindern,
N-0316 Oslo, Norway

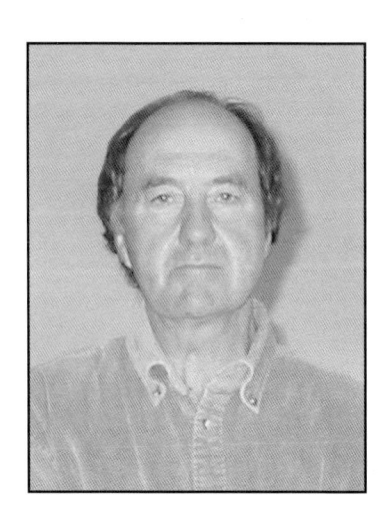

INTRODUCTION

Why phytoplankton?

The phytoplankton provides the main foundation for life in aquatic environments, generating much of the primary food sustaining the rest of the food web in the oceans, lakes, rivers and ponds. The fundamental importance of phytoplankton was recognised early in the development of aquatic sciences, and virtually all of the first institutions established to study oceanography and limnology, in the eighteenth and nineteenth centuries, included phytoplankton among the first parameters to be recorded. As with other fields of natural history, the initial phase of study involved basic observations to catalogue the multitude of life forms found, and develop the taxonomy and systematics needed for establishing consistent records. This revealed an amazing complexity of thousands of species of tiny (microscopic) single-celled organisms, the extent of which is still far from realised.

As with other groups of organisms, the next phase involved monitoring, intended to record the spatial and temporal distribution of the species in order to answer the basic questions necessary to understand their ecology: recording which organisms live where and when, as a first step towards asking important questions of how and why they do so. The basic morphologies of the major groups of phytoplankton are now reasonably well documented, though new species are still consistently encountered, but their ecology is as yet poorly understood due mainly to the failure to establish and maintain long-term plankton records (discussed later).

In the highly technological world of today, the vision of scientists sitting at the microscope laboriously monitoring phytoplankton may seem like very old fashioned science – and as such it has largely been eliminated from national science agendas around the world – but this information is vital for understanding the ultimate threats to human health and well-being posed by the rapid global change of today. Scientists are increasingly being asked to predict the effects of global change in environments with no records to show the effects of previous change. This may be compared to a person trying to work out a sensible travel plan for where to go without first knowing where they are or where they are coming from.

Natural variation and harmful algal blooms

The main thing we know about the temporal distribution of phytoplankton in the world ocean is that it often seems to involve a large amount of variation. Though there are few long-term observations of phytoplankton in the water column (i.e. covering 30 years or more), the spotty record available shows many examples of significant changes in species composition over time-scales of a few days/months/years (e.g. shown by the long-term phytoplankton record at Sherkin Island Marine Station, (Reid, this volume)). Some of this variation is clearly seasonal, but even in strongly seasonal environments it is not usually possible to predict the amounts of various species to be found at any given time. The interplay of factors within the ecosystem causing this dynamic variation is poorly understood, as is the apparent ability of bloom species to bloom periodically while other species seem never to bloom.

One of the main justifications for the initial priority given to phytoplankton studies was the importance of primary producers for sustaining fish stocks. More recently, the public and scientific concern with harmful algal blooms (HABs) has increasingly dominated phytoplankton work. However, this preoccupation with HABs only highlights the more general need to better understand variations in phytoplankton through time. The "sudden appearance" of a harmful species sometimes shocks the public, promoting calls for efforts to find "exceptional" causes (e.g. pollution or transport by ships ballast water), while the well-documented "disappearance" of such species causes problems for maintaining research programs (e.g. reductions in funding) until it reappears. Some but not all HABs are caused by bloom species (e.g. the massive toxic blooms of *Chrysochromulina* in Scandinavian waters in 1988, in contrast to diarrheic shellfish poisoning (DSP), which is caused by low concentrations of species of *Dinophysis* in many regions), and new examples of HABs are regularly discovered that are caused by organisms previously considered unharmful. The only distinguishing feature of HABs is the harm caused to humans, and they therefore should be studied in the broader context of natural phytoplankton variation through time rather than as a special phenomenon.

The major stated goal of HAB research is to understand how HABs occur, in order to predict the timing and extent of their occurrences in future. The scientific community has been successful in generating relatively well-funded HAB projects during the past 30 years, and the extent to which HABs are now understood or predictable may be used as a measure of how successful the pursued research strategies have been. Of particular interest here is that in only one exceptional case, the harmful aerosol-producing HABs of Florida, has long-term monitoring featured as a main element of research strategy. Prediction of HABs has so far generally proved impossible, but the long-term monitoring of phytoplankton in Florida gives a perspective that has provided a notably greater understanding of the phenomena there than elsewhere. More than thirty years of plankton records has allowed the identification of relationships between HABs and climatic cycles (e.g. El Niño), and has suggested novel causes such as the fertilizing of HABs in Florida by atmospheric dust from Africa (Walsh & Steidinger, 2001). We would understand HABs elsewhere better today if appropriate funding similarly had been applied to long-term monitoring in other research programmes.

Environmental micropalaeontology

Some of the main groups of phytoplankton produce micro-fossils (coccoliths, diatoms, and dinoflagellate cysts) that accumulate in bottom sediments, providing a record of changes through time (Figure 1). This record will always be incomplete, due to differential production and preservation of the micro-fossils. However, this can provide supplementary information to the sparse records from the past 100 years of observations from the water column, and is the only information available on timescales of hundreds to thousands of years. Environmental micropalaeontology is thus emerging as a new field of research to extract information on environmental change and biological responses to this from the natural archives of the sedimentary record (Martin, 2000; Haslett, 2002).

THE DINOFLAGELLATE CYST RECORD IN BOTTOM SEDIMENTS

We have studied micro-fossils from one of the major groups of phytoplankton, the dinoflagellates, a group with important primary and heterotrophic production, and including many HAB species. The group is characterised by biflagellated motile cells swimming in the water column, but approximately 10 % of marine species and many freshwater species also produce non-motile resting cysts as part of their sexual cycle. Many cysts are protected by chemically strengthened walls capable of fossilisation, and these empty cysts accumulating in bottom sediments provide a fossil record.

The cyst record in recent sediments representing 1–10s´ of years

Studies comparing cysts in recent sediments (the top 1–2 cm) with plankton records from overlying waters have shown that the cysts in sediments offer a reasonably representative record of accumulated information from the plankton (Dale, 1976). This establishes the use of cysts in sediments as an alternative method for biogeographic surveys to catalogue the spatial distribution of species, for example, where plankton records are inadequate. At least for the cyst-forming species in coastal waters with

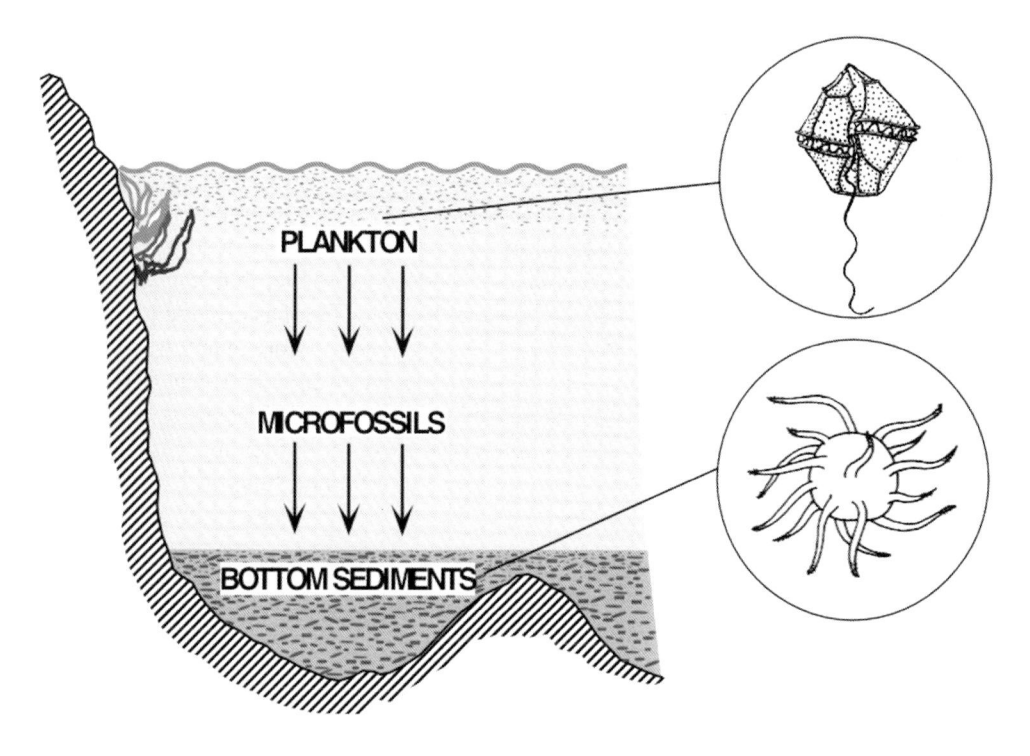

Figure 1: The origin of micro-fossils from the phytoplankton

minimum water and sediment transport, this is arguably easier and more accurate, since a large number of plankton samples would be needed from several water depths and several years, in order to provide information equivalent to that from one surface sediment sample. The biogeographic distributions suggest that at least these cyst-forming species are better environmental indicators than would be suspected from known plankton records (Dale, 1983).

Results from regional cyst surveys often show large variation, with species composition presumably changing in response to variations in environmental factors and ecologic complexity. For example, different cyst assemblages characteristic for river influence, coastal water influence, oceanic influence, and upwelling from the Benguela Current allowed us to identify these aspects of the marine environment off the present day Congo River (Dale *et al.*, 2002). In an ongoing study of the global distribution of dinoflagellate cysts in recent marine sediments, we have treated a database of more than 500 samples with canonical correspondence analysis to produce an ecological classification of more than 200 cyst forms. This provides useful information for interpreting changes in cyst assemblages over various timescales, in successive samples through well-dated sediment cores. Of interest, here, the degree of change seen in the sedimentary record also shows the need for long-term monitoring of the phytoplankton.

The cyst record in cored sediments representing various timescales

10's – 100's of years timescale – Characteristic changes in cyst assemblages allowed us to recognise signals useful for tracing the history of cultural eutrophication in the Oslofjord and industrial pollution in the Frierfjord, southern Norway, during about 150 years starting in the mid 1800s (Dale & Dale, 2002). Ongoing studies are showing clear correlations between maximum eutrophication signals in the cyst record and the timing of collapse of local fisheries at widely different times within the past sixty years along the southern coast of Norway (Dale & Sætre, work in progress). This supports the hypothesis put forward by fisheries biologists that collapse of such fisheries may be connected with eutrophication, and in any case corresponds with marked changes in phytoplankton species composition.

The 1000-year timescale – Changes in cyst assemblages reflect cold/warm climatic cycles within the past 1000 years in the innermost Oslofjord, allowing us to correlate sediments containing valuable archeological remains on land (e.g. Viking ships) with time-equivalent sediments in the adjoining marine basin (to aid extending the archeological work offshore) (Dale & Dale, 2002). These changes correspond to the Medieval warm period (about 1000–1350 A.D.), the Little Ice Age (about 1350–1860 A.D.), and the most recent warming period up to the present day.

Several 1000's of years timescale – Changes in cyst assemblages allowed us to trace the history of "blooms" of a species similar to the toxic *Gymnodinium catenatum* in Scandinavian waters, and show their link to climatic warming around 6,000yrs. BP and 1,000yrs. BP (Dale & Nordberg, 1993). This warmer-water species is almost absent from Scandinavia today, but is found off southern France, northern Spain, and Portugal.

Many 1000's of years timescale – By recording changes in cyst assemblages in successive samples through a 200,000y sediment core taken at a site out from the Congo River, we were able to trace migrations of the Benguela Front in response to global climate change including the extremes of glacial to interglacial modes (Dale & Dale, 2002). This was made possible by using the cyst distributions in recent sediments (noted previously) as a basis for interpreting changes through time.

The main conclusion from the cyst record in sediments

The main conclusion of interest here is that the cyst work supports other phytoplankton records – strongly suggesting marked variation in species composition with time – and extending this to virtually all timescales. The cyst record shows large variations in species composition in response to various natural factors such as climate change. It also shows large variation corresponding to human impact on the environment.

IMPLICATIONS FOR LONG-TERM MONITORING OF PHYTOPLANKTON

The cysts provide a restricted view, produced by only about 10% of the marine species, but there is no reason to suspect that cyst-forming species are unrepresentative of the ecological variation within the group. The fact that records such as those from dinoflagellate cysts show so much variation, both spatially and temporally, is almost certainly just a brief glimpse of a much more comprehensive variation in the phytoplankton as a whole. In this respect, the cyst record may be regarded as a more easily obtained reminder of potentially large amounts of variation in species composition within the phytoplankton, information that is missed without adequate sample coverage in space and time through long-term monitoring.

HAB research

If, as concluded here, HABs are basically part of the natural phytoplankton variation that may or may not be affected by human impact on the environment, one of the best ways to study them must be from observations of their natural occurrences (which cannot be reproduced in the laboratory). However, since HABs often are episodic, there is an obvious need for long-term observations. Furthermore, since there is no evidence for harmful species behaving especially differently from others (which in any case may prove to be harmful eventually), there is a need to monitor the *whole* phytoplankton as thoroughly as possible.

Long-term data sets for HABs – status

There are almost no long-term data sets available – the one notable exception being *Karenia brevis* in Florida and the Gulf of Mexico noted above. New HABs are therefore usually recorded with little/no long-term background ecological setting established, and with emphasis on harmful species only. HAB information is therefore being accumulated in a fragmented form that makes it difficult/impossible to synthesise into a bigger concept of the phenomenon (many small bricks with no mortar with which to build). This represents a major barrier to future progress.

The need for long-term monitoring

Long-term monitoring is the only way of recording episodic data on the necessary timescales. Furthermore, documenting where and when such phenomena occur is still a sound way of approaching the important questions of how and why they occur. This may be seen by considering how physical and chemical oceanography have benefited greatly from 20–30 year records (the discovery by Sy *et al.* (1997); of rapid spreading of large amounts of intermediate waters across the North Atlantic or understanding the extent of El Niño). Long-term monitoring would provide the necessary framework for better assessing the results of many of the individual HAB projects (e.g. as reported at the international HAB meetings).

If long-term monitoring is so important then why is it largely ignored?

This has to be seen as part of a general *trend* in The Natural Sciences – where the "natural" aspects receive less attention as "science" develops. Direct observations of the natural environment such as plankton counting is perceived by some as "old fashioned" while that which is considered to be modern science becomes increasingly less distinguishable from technology. Technological developments in HAB research, such as chemical and genetic methods for identifying toxins and strains of organisms, are at best aids for studying the basic phenomenon. Unfortunately, in the tight competition for research funding, the more basic observational research on HABs (e.g. long-term monitoring) is often perceived as "old-fashioned science" and loses out.

It should be noted that this *trend* is not easy to reverse. Accurate identification of organisms is still crucial to both natural observations and culture work in the lab, but there is a danger of lowering the status of traditional taxonomic work in favour of "biomarkers", which so far do not seem likely to adequately fill the taxonomic void once created. If this negative trend continues, phytoplankton biology will be faced with a major crisis in just a few years from now. Almost all of the expertise for identifying phytoplankton is currently held by a relatively small group of scientists, mostly soon approaching retirement age. If their expertise is not passed on, the next generation of scientists interested in considering the phytoplankton risk literally not knowing what they are talking about.

It is appropriate, here, to acknowledge the scientific foresight of Matt Murphy, Director of Sherkin Island Marine Station. As one of the first to recognise the need to reverse this negative trend in phytoplankton research, he started the ongoing long-term monitoring programme at the station in 1978 (see Reid, this volume), and among many other great efforts has supported the work of one of the true masters of phytoplankton taxonomy, Enrique Balech, in Argentina (Balech, 1995).

The case for establishing a global marine phytoplankton monitoring program

There is growing awareness of the need for global integrated efforts to understand the basic systems of the planet. Within this framework, a strong case can be made for global monitoring of marine phytoplankton – accounting for about half of the total primary production, and including the problems of HABs. Not least, there will be an increasing need for "State of the environment" monitoring of the impact of the rapidly growing human population (the HAB research community is already responsible for the widely held belief that there is a global epidemic of HABs caused by cultural eutrophication, first suggested by Smayda, 1990). The establishment of global monitoring stations is not reliant on heavy increases in funding. Substantial research efforts in marine phytoplankton are being funded in many parts of the world – it is a case of committing a reasonable percentage of present day budgets to a *long-term* international effort (the Ocean Drilling Program in geology is a good example of such cooperation).

The main requirement is to standardise methods and taxonomy (including links to genetic information), so that results may be integrated globally and quantified in order that modern statistical treatments can be applied. Much of this could be achieved through international workshops. Such a global effort would quickly produce relevant information about blooms (including HABs), from stations covering different climatic zones and different degrees of human impact; in the long-term, it would provide the all-important data covering periodicity. This contrasts with many efforts today which may be characterised as short-term efforts to study all-too-often illusive harmful blooms.

CONCLUSIONS

The sedimentary record of dinoflagellate cysts shows significant changes in species composition on all timescales. Even so, this incomplete view, produced by only about 10% of the marine species, is almost certainly just a brief glimpse of a much more comprehensive variation in the phytoplankton as a whole. Such critical information regarding phytoplankton variation is missed without adequate sample coverage in space and time through long-term monitoring.

There is a strong case for suggesting that important information of how/why blooms (including HABs) occur will only be revealed through long-term monitoring – in which case it becomes a question of to what extent we can claim to be studying bloom phenomena *without* long-term monitoring?

The time is ripe for considering where we are/are not going with HAB research. In my opinion:

- There is a need to shift emphasis from technological developments back to basics.
- Real efforts should be made to re-establish the status of monitoring.

The possibility of establishing a long-term global phytoplankton-monitoring programme should be seriously considered. This would offer one of the few possibilities for following future changes in this fundamentally important part of the biosphere, in response to global change from natural climatic cycles and the environmental impact of a rapidly growing human population.

Barrie Dale *is Professor in the Dept. of Geosciences, University of Oslo, Norway Education: minimal – but includes PhD in Geology, The Open University, UK. Previous Employment:*

1958–1962 Research Technician, Geology Dept., Sheffield University, UK

1962–1963 UN International Work Camp worker – Greece

1964–1975 Research Associate, Woods Hole Oceanographic Institution, USA.

His recent employ (1975–present) has been in Oslo. Prof. Dale's teaching experience includes: Director of Studies – 3-year BSc programme in Environmental Sciences and teaching environmental sciences and micropaleontology at undergraduate and graduate levels. His research activities focus on the use of microfossils in sediments as environmental indicators – living/recent/fossil – eg to provide long-term records of environmental change covering pre-historic to present-day. He has a special interest in natural variation versus human impact (climate/marine pollution/harmful algal blooms/overfishing), and has co-operated with the European Commission on Harmful Algal Blooms and sustainable development. In 2004 he received the American Assoc. of Stratigraphic Palynologists Medal for Scientific Excellence.

3.11

ROCKY-SHORE MONITORING AT SHERKIN ISLAND MARINE STATION SINCE 1975

Gillian Bishop
Environmental Consultant,
Loan of Durno, Pitcaple,
Aberdeenshire, Scotland,
UK

INTRODUCTION

Why Sherkin? Reasons for establishing the monitoring programme

The rocky shore is a complex and dynamic habitat, influenced by many environmental factors. The rocks themselves vary from being hard, soft, smooth or creviced, thus offering different surfaces that suit different combinations of animals and plants. Water temperature, air temperature, wind strength, wind direction and rainfall all play their part in determining which species colonise, survive, thrive and reproduce successfully on the shore, with biological interaction being the final, important, structural component which shapes the resultant plant and animal community.

In October 1974 the oil tanker *Universe Leader* spilled approximately 2,600 tonnes of Kuwait crude oil into the waters of Bantry Bay. Three months later, in January 1975, a second spill of about 460 tonnes of heavy fuel oil occurred. Seashore monitoring sites, located and described by Crapp (1973), had already been established in Bantry Bay shortly after the oil transshipment facility opened on Whiddy Island in 1969. These sites were re-surveyed following the two accidents in order to establish whether any damage had occurred to the intertidal flora and fauna (Baker *et al.*, 1981).

At Sherkin these incidents focused attention on the fact that the clear Atlantic waters of Long Island Bay and Roaringwater Bay had no similar industrial threat hanging over them and as such should be described and studied in order to record naturally occurring changes in the marine plant and animal communities. And so, in summer 1975, the first transects were established and surveyed on Sherkin and several other islands and mainland shores within Roaringwater Bay. By 1981 the survey had developed meticulous and robust methodology. This and additional rocky shore surveys have developed and grown geographically and continue to this day.

Annual Sites

The annual sites fall into five natural areas: Roaringwater Bay (Figure 1, area 1), Dunmanus & Bantry Bays (Figure 1, area 2), East of Baltimore (Figure 1, area 3) and Cork Harbour (Figure 1, area 4); with the number of sites in each area reflecting the differing complexities of their coastlines.

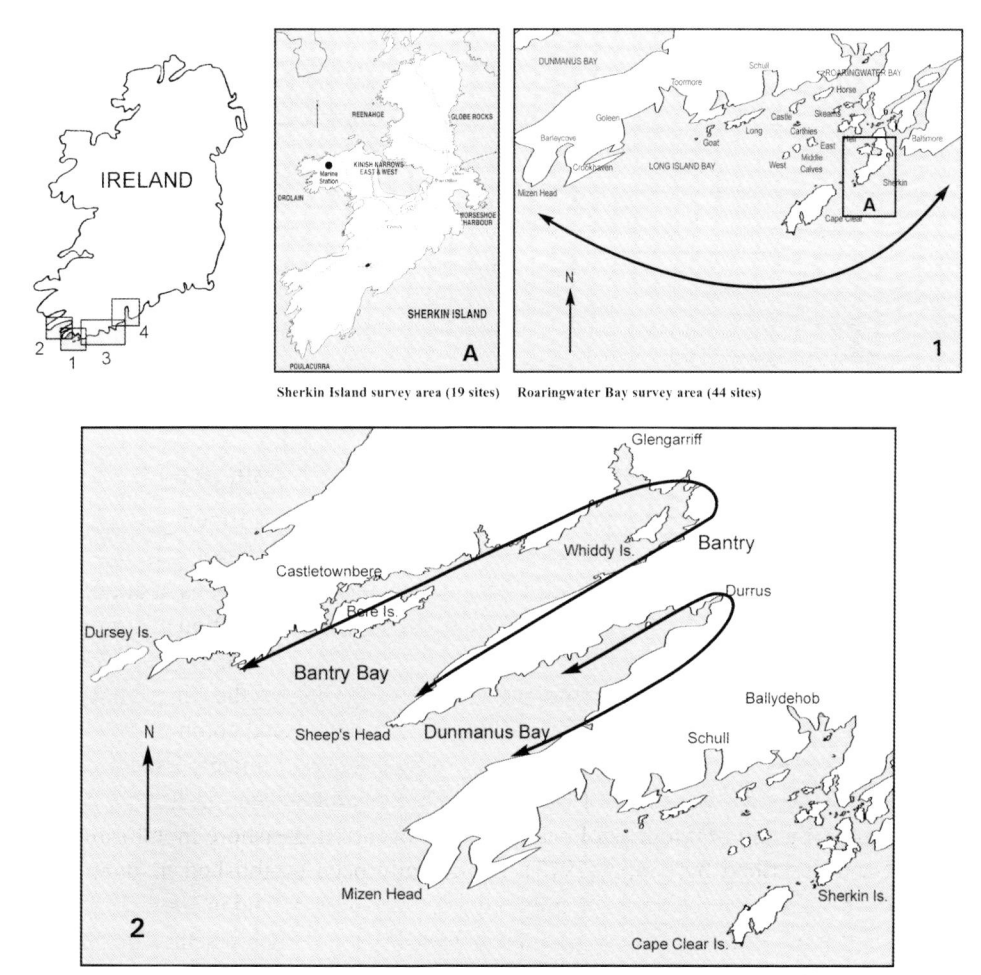

Sherkin Island survey area (19 sites)   Roaringwater Bay survey area (44 sites)

Bantry Bay (18 sites) and Dunmanus Bay (8 sites) survey areas

Figure 1: Sherkin Island Marine Station Rocky Shore Survey Programme

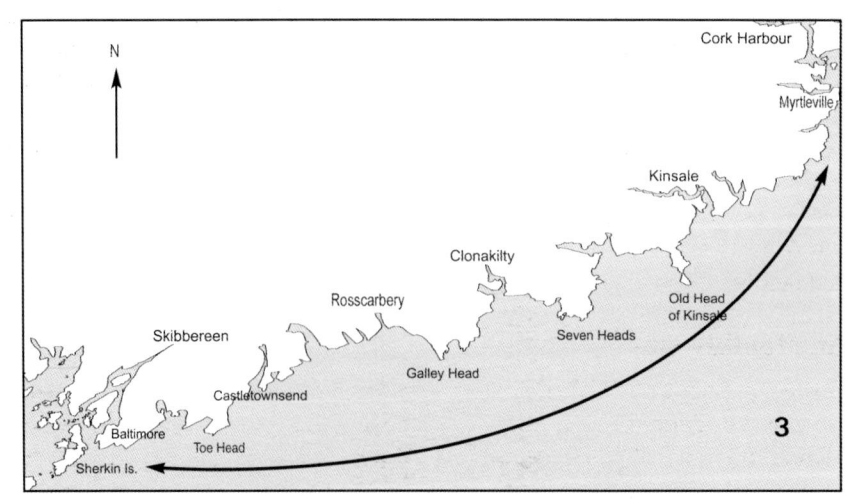

East of Baltimore survey area (30 sites)

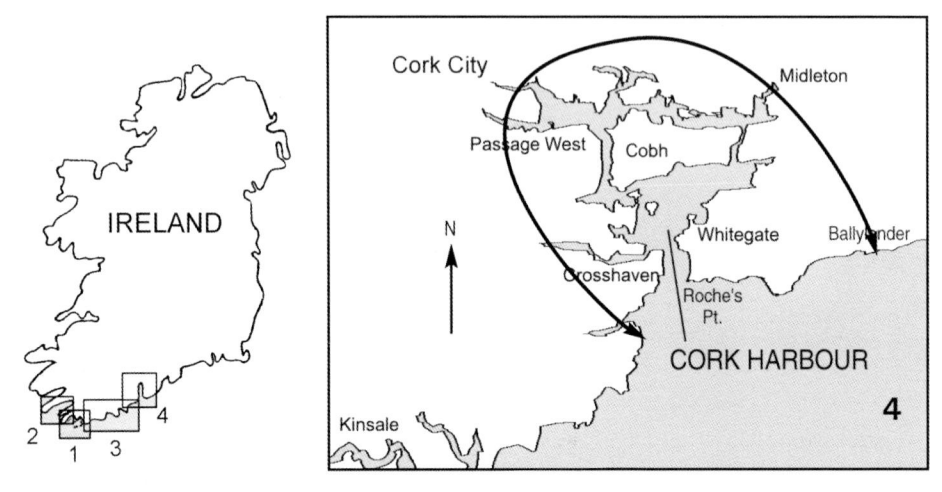

Cork Harbour survey area (27 sites)

Figure 1 continued

Roaringwater Bay has 62 sites that are surveyed annually. These include 19 on Sherkin Island, 29 on the various islands and 14 on the mainland coast around the Bay. Data have been collected since 1975 and are reported in Hunter (1997), Egerton (1998), Falcous et al. (2001a), Jackson et al. (2001) and Migus & Moore (2002).

Eight sites in Dunmanus Bay were added to the programme in 1981 (Layden & Lane, 1984; Taylor & Hunter, 1996; Clark et al., 1997; Egerton et al., 1998; Ratcliff, 1999; Browne et al., 2001a).

Bantry Bay has 18 annual sites, the majority of which were set up in 1995 (Taylor & Hunter, 1996; Clark et al., 1997; Egerton et al., 1998; Clark 1999; Edwards et al., 2001).

Sites along the mainland coast to the east of Sherkin were set up in 1995. The East of Baltimore survey comprises 30 rocky shore sites between Baltimore and the outer edge of Cork Harbour (Taylor & Hunter, 1996; Clark *et al.*, 1997; Egerton *et al.*, 1998; Clark, 1999; Browne *et al.*, 2001b). Within Cork Harbour, 27 sites (Taylor & Hunter, 1996; Clark *et al.*, 1997; Egerton *et al.*, 1998; Bolger-Hynes 1999; Clark 1999; Falcous *et al.*, 2001b) complete the annual programme.

Each rocky shore is sampled at the same time of year within a two-week window.

Methodology is strictly standardised and based on the monthly survey techniques described below.

Sherkin Monthly sites

Seven sites on Sherkin Island (Figure 2) were chosen to reflect a range of aspect and exposure to wave action. They comprise a very exposed shore (Poulacurra), an exposed shore (Drolain Point), three moderately exposed shores (Globe Rocks, Horseshoe Harbour and Reenahoe) and two sheltered shores (Kinish Narrows East and Kinish Narrows West). The two sites at Kinish Narrows (East & West) lie 50m apart either side of the entrance to the shallow marine inlet of Kinish Harbour. Both are sheltered, with similar geology and similar physical profiles, but have different aspects. Waves breaking directly onto Kinish Narrows West are from a northerly direction and have only crossed the Heir Island/Sherkin Island channel, whereas waves breaking onto Kinish Narrows East arrive from the northwest and have more energy, having travelled further from the line of islands in the centre of Roaringwater Bay.

Methods

The sites are surveyed monthly, during the growing season, between April and October by recording species abundance along a single transect per shore. Abundance is recorded as either percentage cover (algae and ground cover animals such as mussels and barnacles) or number per quadrat (most animal species). Detailed maps and photographs together with permanent markings enable precise relocation and sampling of a transect perpendicular to the water's edge from EHWST to ELWST (extreme high and low water spring tides respectively). Recording is carried out within two laterally contiguous $0.25m^2$ quadrats placed at 30cm vertical intervals i.e. one tenth of the tidal range (Nelson-Smith, 1967). Resultant stations number between nine and fourteen per transect. The area inside each quadrat is viewed as a two-dimensional plane, thus deep crevices, rock pools and undersides of boulders or stones are not included in the records. Competency of biologists with regard to identification skills and understanding of methodology is verified at the start of each survey season in order to assure consistency in the rocky shore programme.

Results

Species abundance and density figures for each quadrat are added monthly to the Station's seashore database which has been interrogated to enable descriptions of species richness, spatial distribution of species down the shore, seasonal and annual species variation to be made. A comprehensive description of the shores has recently been published (Bishop, 2003) and key points emerging from the survey programme of interest to the current discussion of long-term monitoring are addressed in this paper.

With such a wealth of accumulated data, analysis has, so far, concentrated on the dominant species within each functional group on the shore. This means looking at the animal species that comprise ground cover (barnacles and mussels); two or three algae of the understorey and of the canopy layer; two or three grazing animals (periwinkles, topshells or limpets) and the most common invertebrate predator of the shore – the dogwhelk. For each chosen species (selected species differ between shores), its distribution down the shore, monthly and annual variations in density have been investigated using data from 1981 to 2000.

Figure 2: Monthly sampling sites on Sherkin Island

Species Richness (Figure 3) ranged from a maximum of 268 taxa at Horseshoe

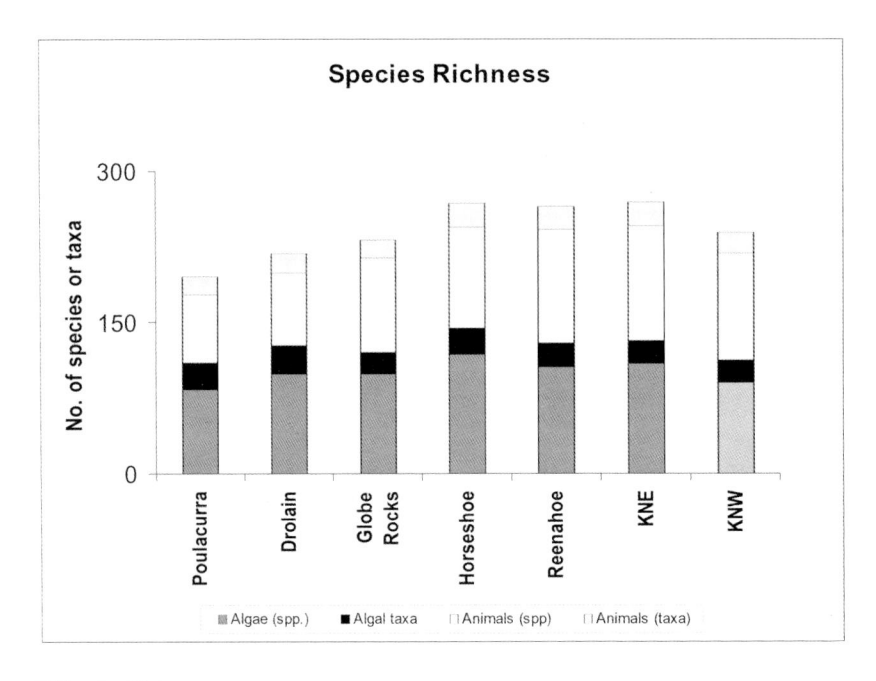

Figure 3: Species Richness

Harbour to a minimum of 196 at Poulacurra. There was very little difference in the number of taxa at the moderately sheltered sites, but the very exposed and exposed sites were less rich and had more plant than animal species. Of the three moderately exposed sites, Globe Rocks shows less diversity in both plants and animals than Horseshoe Harbour and Reenahoe; and of the two Kinish Harbour sites, the west appears to be less diverse with a smaller number of algal species.

Seasonal Variation is evident in certain algae and a few of the dominant animal species. The ephemeral green algae *Enteromorpha* spp., *Ulva lactuca* and *Cladophora rupestris* showed succession in their appearance at Drolain Point throughout the spring and summer with peaks occurring in April (*Ulva lactuca*), June (*Cladophora rupestris*) and July (*Enteromorpha* spp.). Patterns of *Enteromorpha* spp. variation from April to October appear to be very similar regardless of wave exposure as shown by variation at three sites – Drolain Point, Reenahoe and Kinish Narrows East (Figure 4). Red algae showed seasonality at certain locations. For example pepper dulse (*Osmundea pinnatifida*) decreased steadily through the summer at Horseshoe Harbour, but this was not recorded at the other moderately exposed sites Globe Rocks or Reenahoe. *Lomentaria articulata* also declined from April to October at two of these sites

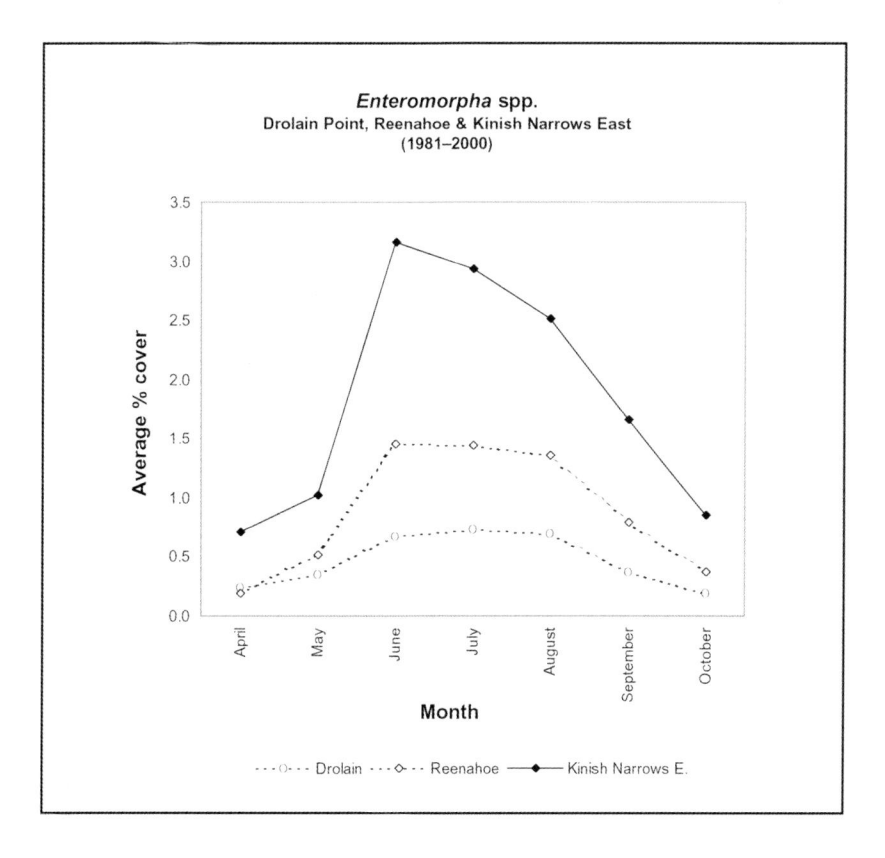

Figure 4: Seasonal variation in *Enteromorpha*

(Horseshoe Harbour and Reenahoe). Among the canopy layer, evidence of seasonality is sparse.

Although not so common in animal species, seasonal variation is occasionally found and is shown here in the flat periwinkle (*Littorina obtusata*) and barnacles. At four of the five sites where *L. obtusata* was recorded, an August peak in density was followed by a September decline in numbers, while at the fifth site a peak in numbers occurred during July followed by a decline in August (Figure 5). Barnacles at all seven sites showed a steady overall increase in cover from April to October (Figure 6), but whereas the rate generally levels off at the end of the summer, at two sites (Kinish Narrows East and Globe Rocks) cover increased markedly during this period.

Annual variation in population size is evident in several of the large brown algae, barnacles, mussels, limpets and dogwhelks.

Many of the large brown algae declined during the twenty year period 1981 – 2000 (Figure 7), with more decline seen during the 1990s than the 1980s. Four exceptions to this were *Fucus vesiculosus* and *F. spiralis* at two sites – Globe Rocks and Reenahoe. Large variations in density of animal ground cover species have been found. A major change in barnacle abundance occurred at all seven sites in 1991 (Figure 8). Population

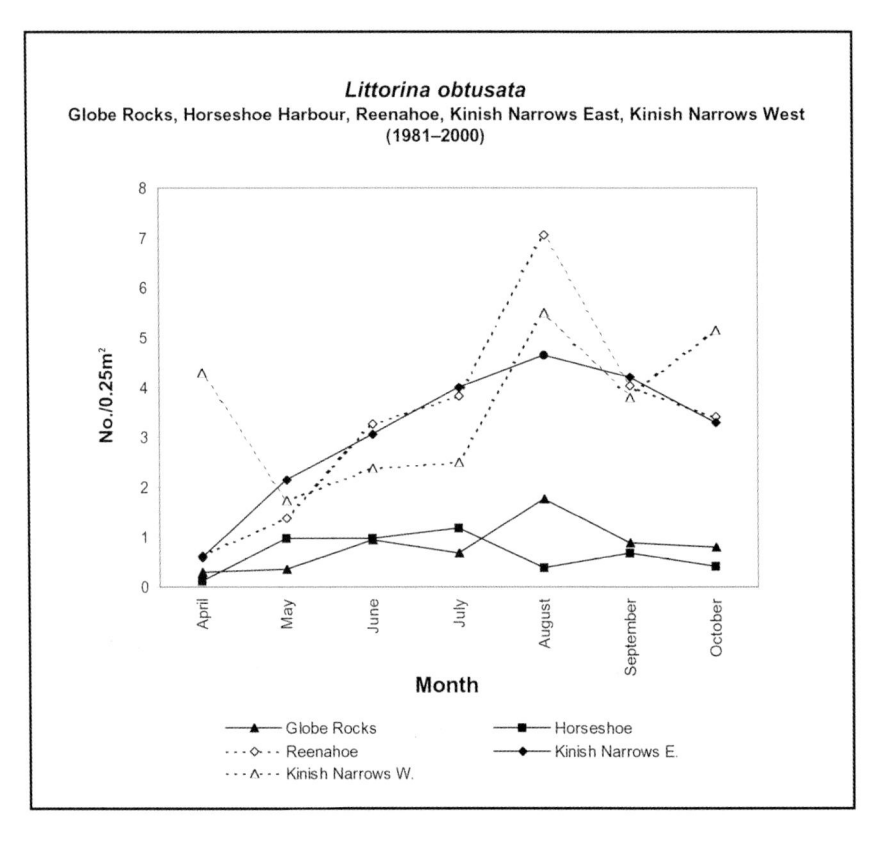

Figure 5: Seasonal Variation in the flat periwinkle *Littorina obtusata*

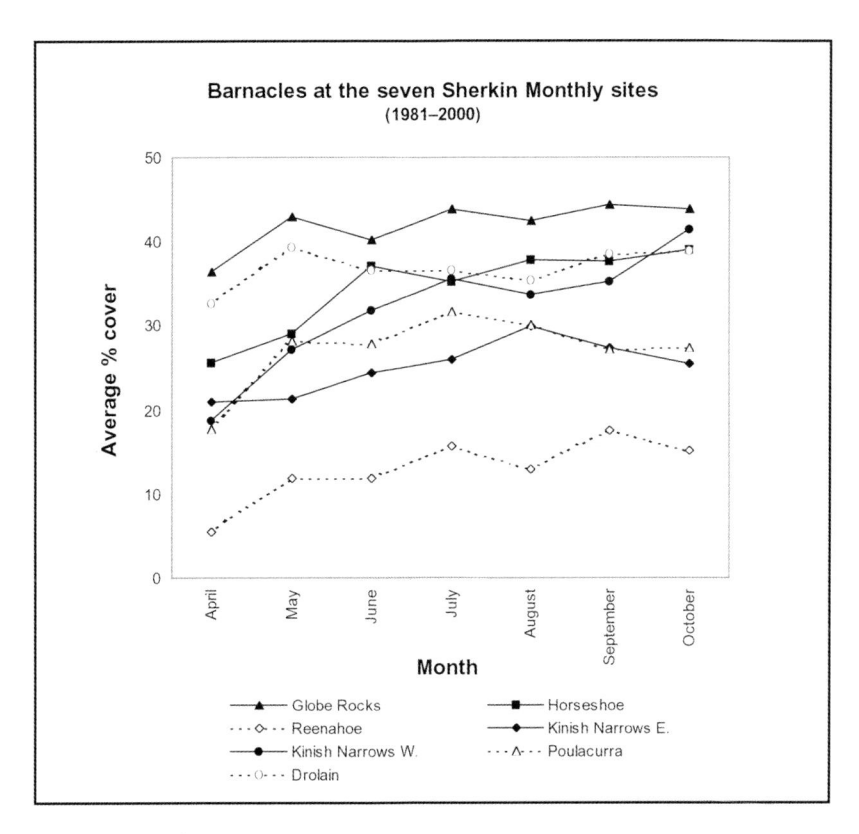

Figure 6: Seasonal Variation in barnacles

densities were generally high and increasing throughout the 1980s, but a universal crash in numbers was observed during 1991. Since then, there has been a certain amount of recovery, more rapid at some sites e.g. Kinish Narrows West, than others e.g. Kinish Narrows East. At the very exposed site Poulacurra, the population decline of 1991 was less severe but definitely present. Similarly, mussels where present in any quantity (on the exposed shores of Poulacurra and Drolain Point), were less common in the 1990s than the 1980s, with 1991 recording the lowest densities at both sites (Figure 9). At Drolain recovery has been less successful than at Poulacurra, although a marked increase has occurred during 1999 and 2000. Numbers of grazing limpets have been generally consistent over the twenty year period (Figure 10), although population variation at Horseshoe Harbour has been conspicuously different and somewhat erratic with an overall trend of increase.

The predatory dogwhelk is found on all seven sites, but population sizes and rates of change have varied between sites (Figure 11). Throughout the survey period populations have been consistently larger at three sites: Poulacurra, Drolain Point and Horseshoe Harbour, but various similarities and differences between populations at all

	Pelvetia canaliculata	Fucus spiralis	Fucus vesiculosus	Fucus serratus	Ascophyllum nodosum	Himanthalia elongata
Poulacurra						Absent after 1989.
Drolain Point	Thicker in 1980s than 1990s.			Increasing during 1980s. Absent or sparse from 1993–2000 (except 1998).		Absent or sparse 1993–2000.
Globe Rocks	Declined throughout 20 years. Very sparse from 1994.	Generally sparse in 1980s. More common in 1990s.	Sparse until 1997. Fivefold increase between 1997 and 2000.	Declined 1984–1993.		
Horseshoe Harbour		Declined 1984–2000.	Declined 1984–2000.	Declined in 1980s. Remained sparse in 1990s.		Declined in 1990s.
Reenahoe		Increased 1985–1990. Virtually absent 1991–1997.	Increased in late 1980s. Dominant wrack in 1990s in middle shore.	Declined slightly over 20 years.	Dominant on middle shore in early 1980's. Declined and low 1986–2000.	
Kinish Narrows East		More common in late 1990's than earlier.	Less common 1988–1996 than earlier and later.	Declined in 1990s.	Declined in 1990s.	
Kinish Narrows West		Generally low abundance. Isolated peaks 1993 and 1996.	Declined 1991–2000.	Declined slightly in late 1990s.	Declined in 1990s.	

Figure 7: Annual variation – major changes in the large brown algae

Figure 8: Annual variation in barnacles

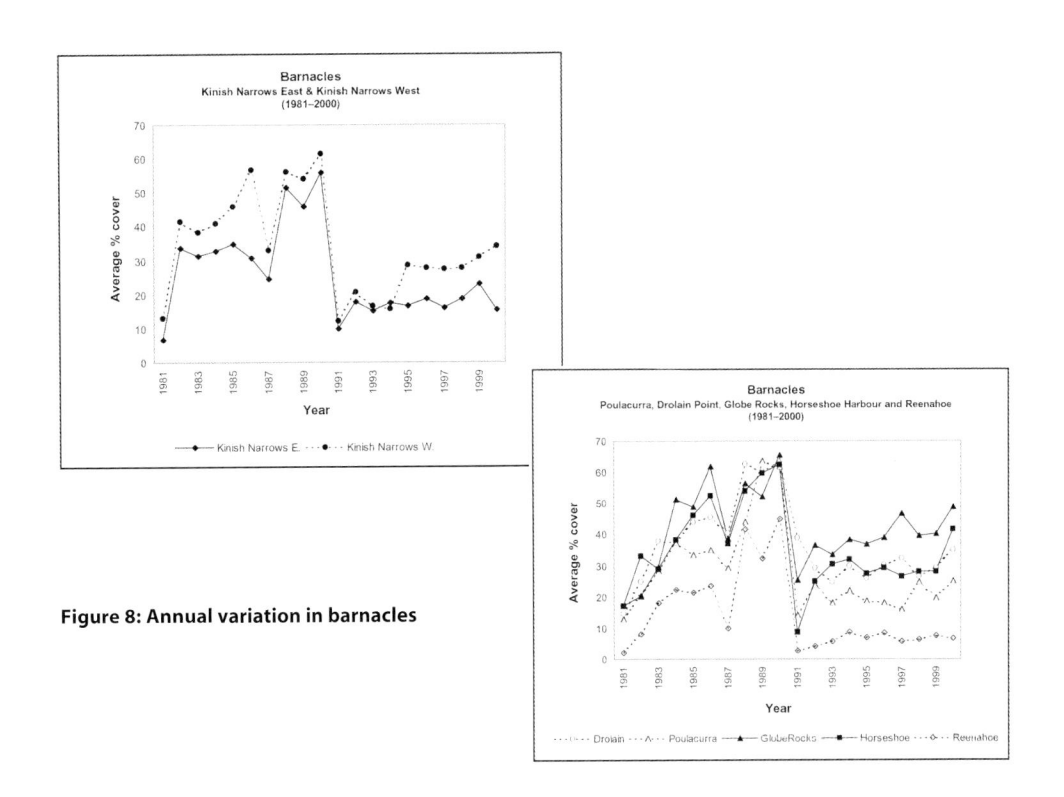

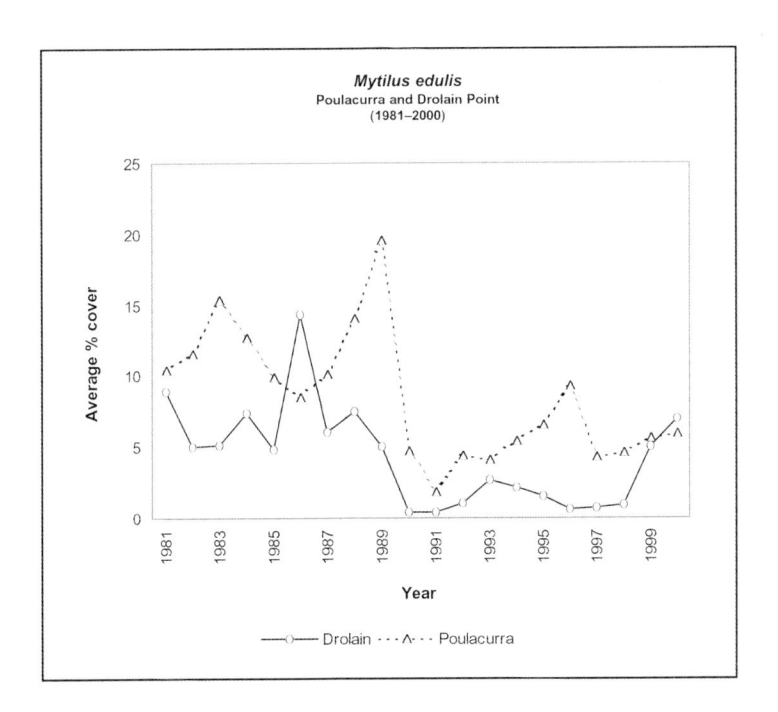

Figure 9: Annual variation in mussels

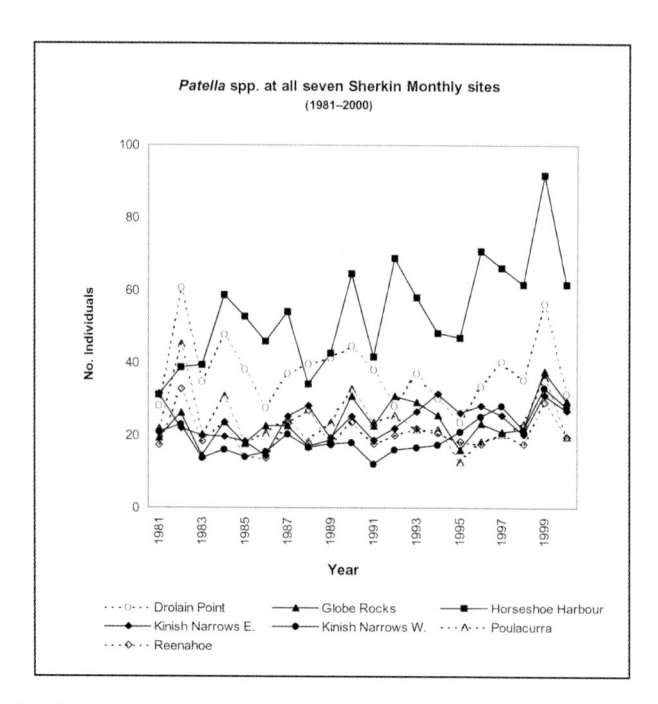

Figure 10: Annual variation in limpets

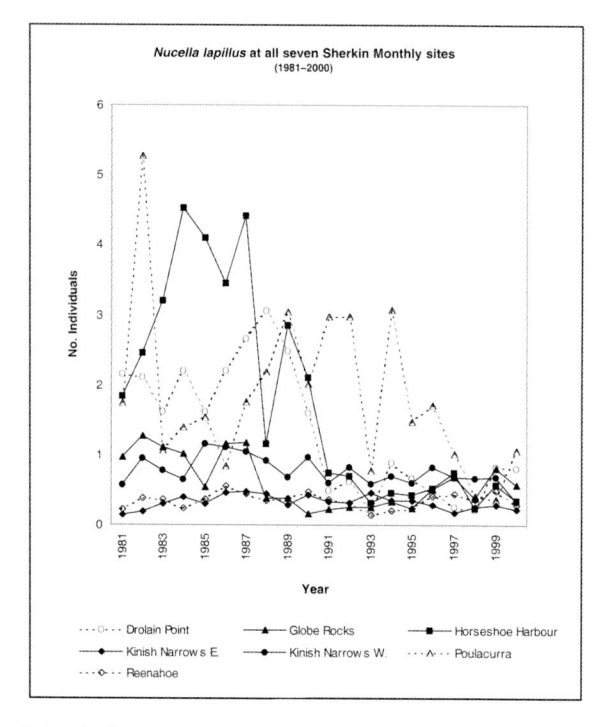

Figure 11: Annual variation in the dogwhelk *Nucella lapillus*

seven sites have been found regardless of overall population size. At five of the sites dogwhelk numbers were consistently higher during the 1980s than the 1990s, with the main decrease occurring during 1991. This is especially obvious in the larger populations of Drolain Point and Horseshoe Harbour. At Poulacurra, a population decline of equal magnitude was recorded, but not until 1995. Between 1992 and 1993, numbers were halved, but recovered in 1994; however the decrease in 1995 continued to 2000 with no return to the higher numbers of the late or early eighties (3 and 5 per quadrat respectively). Dogwhelk numbers at Reenahoe performed differently with a population increase occurring between 1993 and 1999 after the minimum population size recorded in 1993.

Relationships between species

Since a major influence on species success is quantity and quality of food supply, certain groups of species were analysed together. The algae-grazer relationship was investigated by looking at the combined performance of fucoids and the flat periwinkle at Horseshoe Harbour and that of fucoids and the limpet (*Patella* spp.) at Reenahoe (Figure 12). At Horseshoe Harbour both populations decreased in the 1980s and were minimal from 1989 onwards apart from a large single increase in winkle numbers in

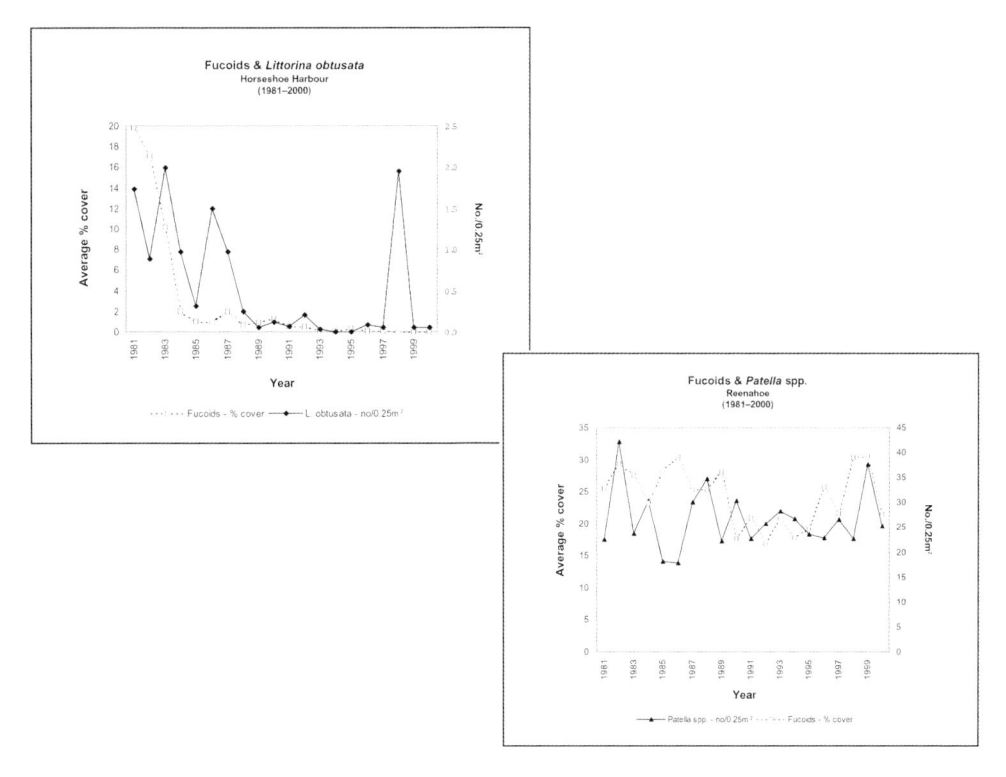

Figure 12: Annual Variation in algae and grazers

1998. The decline in fucoid cover (1992, 1993 and 1994) generally preceded the fall in winkles (1994, 1995), and while fucoids remained sparse, various population increases occurred in the winkles (1983, 1986 and 1988). If these increases reflect juvenile recruitment, it appears that there is insufficient fucoid habitat to support them. The data also suggest that even the minimal winkle population size present during 1989 through to 1997 prevented re-establishment of a rich fucoid canopy due to the grazing pressure on the algal sporelings. Limpets also feed on sporeling algae and scrape diatoms from rock surfaces. Their grazing can also substantially weaken the stipes of fucoids. At Reenahoe there is a direct but offset relationship between fucoids and limpets. Fucoid density was high in six different periods during five of which limpets decreased in number. When limpet numbers increased, fucoid cover decreased in four of the seven recorded periods; and it can be seen that the periods of limpet population increase occurred one or two years following fucoid increase. Over the twenty year period, larger increases in the limpet population were associated with fucoid cover greater than 25%; below this figure increase in limpet numbers were not so marked and the population size remained between 22 and 30 per quadrat.

Predator-prey relationships were analysed by reference to barnacle and dogwhelk data from Globe Rocks and Reenahoe (Figure 13). Early in the survey at Globe Rocks barnacles increased during 1982–1986 and dogwhelk numbers fell within this period (1983–1985). In 1986 both populations increased. Following this, barnacle cover reduced in 1987 and dogwhelk numbers declined a year later (1987–1988). For the next eight years, 1988–1995, dogwhelk numbers remained low, while barnacles increased from 1987 to 1990 only to crash dramatically in 1991. There is no evidence to attribute this crash to grazing pressure, since dogwhelk numbers had fallen and remained low for three years consecutively prior to the major reduction in barnacle cover.

The barnacle population recovered at a steady rate throughout the 1990s, and it appears that the population size achieved by 1996 was sufficient to support the marked increase in dogwhelk numbers seen in 1986 and 1987. Both populations fell in 1998, but the level of barnacle cover was still sufficient to support a further increase in dogwhelk numbers the following year. At Reenahoe, the pattern of predator-prey variation is broadly similar. When barnacle cover fell abruptly from 1986 to 1987, there was a decrease in dogwhelks the same and subsequent year. This was also evident during the large crash of the barnacle population in 1991; dogwhelks declined in 1991, 1992 and 1993. As well as this pattern of delayed decline of the predator following prey reduction, long periods of combined increase have also occurred. During the six year period 1981–1986 when barnacle cover was continually increasing, dogwhelks also showed an overall increase in numbers despite a single year of reduction in 1984.

At Reenahoe and Globe Rocks the predator-prey relationship is relatively simple, but where mussels form a significant part of the ground cover, the relationship is more complex as at Poulacurra and Drolain Point (Figure 14). Dogwhelks are known to prefer mussels as a food species. At Drolain Point all three species were more abundant during the 1980s than the 1990s. However the timings of increases, decreases and recovery of each species varied. Mussels began to decline in 1987, while barnacle and dogwhelk populations continued to increase. This was well before the decline of

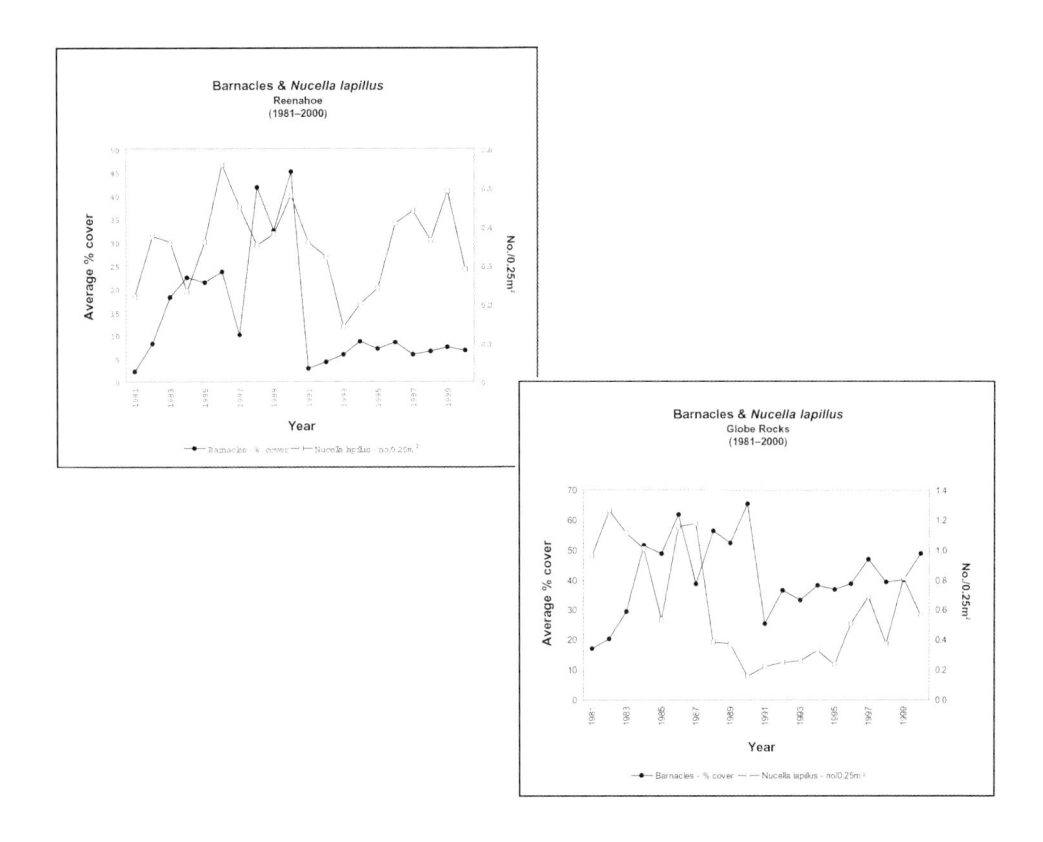

Figure 13: Predator-prey relationship at Globe Rocks and Reenahoe

dogwhelks (1989–1993) and of barnacles (1991–1993). It seems likely that the pressure on barnacles from dogwhelk feeding increased as mussels began to decline. At Poulacurra, patterns of interrelated species variation are not so clear. The sizes of the barnacle and mussel populations at the start of the survey period were evidently insufficient to sustain the large increase in dogwhelk numbers in 1982, but as both prey populations increased throughout the decade, so too did the dogwhelks from 1984 to 1989. Rates of increase of both prey species were more rapid in the early and late 1980s than in the period 1984–1986; during 1987–1990 when barnacles occupied 30–40% of the shore, the rate of increase of all three species was at its highest. During the 1990s after the barnacle crash, although mussels increased at a slow and steady rate, numbers of dogwhelks fluctuated for three years, but then repeatedly decreased.

Patterns of Species Abundance

Many of the dominant functional species on the shore have shown major differences in levels of abundance over the twenty year period and these have fallen naturally into a comparison of their performance in the 1980s and 1990s (Figure 15). While a few have been more successful in the 1990s at certain sites (*Fucus spiralis, F. vesiculosus,*

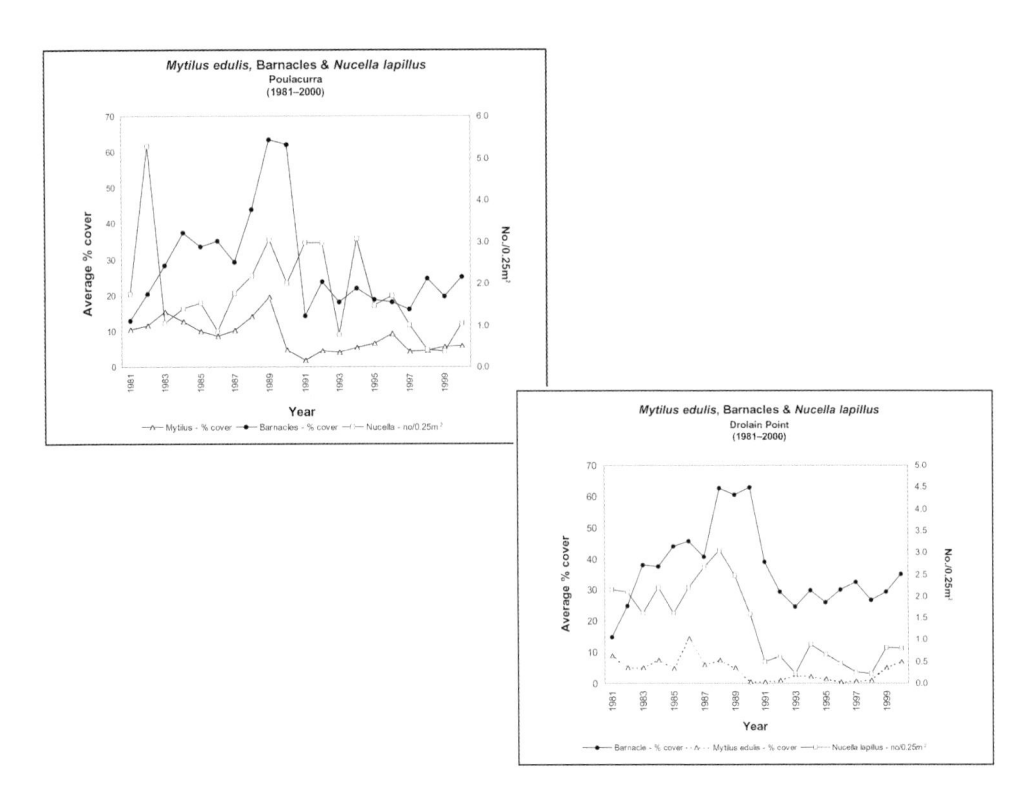

Figure 14: Predator-Prey Relationships at Poulacurra and Drolain Point

Littorina neritoides, Gibbula umbilicalis), the majority have performed much better throughout the 1980s. The overall conclusion therefore has to be that productivity on Sherkin's rocky shores has declined in the 1990s. However as recording began in 1981 only, we have no way of knowing whether the population levels of the 1980s were 'normal' or whether they were particularly high. It is equally possible that the low population densities of the 1990s represent the true viability of Sherkin's rocky shore environment.

Within this major 'divide' of the two decades, several species also show shorter cycles of abundance (Figure 16). What is interesting is that these cycles are inconsistent both within and between shores. Abundance of bladder wrack at Reenahoe for example can be divided into cycles lasting 4, 7, 6 and 6 years; whereas the same species at Kinish Narrows East shows cycles lasting 8, 7 and 6 years. The length of cycles of abundance of the small periwinkle range from 3 years at Globe Rocks to 10 years at Drolain Point. Thus no two shores have the same cycle length for any particular species and cycles are not a feature of every shore.

	Poulacurra	Drolain	Globe Rocks	Horseshoe Harbour	Reenahoe	Kinish Narrows East	Kinish Narrows West
Pelvetia canaliculata	✓	✓					
Fucus spiralis						✓ (1990s)	
Fucus vesiculosus			✓ (1990s)	✓			✓
Fucus serratus		✓	✓	✓		✓	
Himanthalia elongata	✓	✓		✓			
Ascophyllum nodosum					✓	✓	✓
Laminaria digitata	✓						
Mastocarpus stellatus	✓	✓	✓	✓			
Osmundea pinnatifida				✓			
Enteromorpha spp.							
Barnacles	✓	✓		✓	✓	✓	✓
Mytilus edulis	✓	✓					
Littorina obtusata				✓			
Melarhaphe neritoides	✓ (1990s)			✓ (1990s)			
Littorina littorea							✓
Gibbula umbilicalis					✓ (1990s)		
Nucella lapillus		✓	✓	✓			✓

✓ higher in 1980s
✓ (1990s) higher in 1990s

Figure 15: Major difference between species abundance levels in the 1980s and 1990s

Species	No. of years of cycle of abundance						
	Poulacurra	Drolain Point	Globe Rocks	Horseshoe Harbour	Reenahoe	Kinish Narrows East	Kinish Narrows West
Fucus vesiculosus					4, 7, 6, 6	8, 7, 6	
Fucus serratus					7, 6, 6, 4		
Littorina obtusata	-	-	8,5,7		6,4,5,6	9,5,6	3,6,11
Littorina littorea					6,4, 7, 4, 2	4, 8, 6, 4	
Melarhaphe neritoides	-	7,10	6,3,3,4,6,3				
Osmundea pinnatifida	6,8, 8	-	-		5, 6, 10		
Enteromorpha spp.		6, 8, 6				4, 7, 9	
Mytilus edulis	6, 6, 7, 4						
Barnacles	7, 5, 10		7,5,10				
Patella spp.				8, 4, 5, 6			

Figure 16: Length of Species Cycles of abundance within the twenty year period (1981–2000)

DISCUSSION

Key Findings: What have we learned since 1975?

There are key findings in both an ecological and methodological sense. Seven ecological points are clear:

- We have recorded real change in biological populations which has been incremental in some years and larger in others;
- There was a widespread reduction in barnacle populations during 1991;
- Abundance of many species was very different in the 1980s when compared to the 1990s;
- General productivity on Sherkin's rocky shores was greatly reduced during the 1990s;
- Species diversity has remained constant;
- Some species display definite cycles of abundance;
- Cycles of species abundance are not necessarily consistent within or between sites.

From a methodological stand, several points are pertinent to a discussion of long-term monitoring. During the early years of the survey programme, methods were continually improved to enable the collection of data by a changing team of biologists to be both rigorous, consistent and of a high standard. Neither the biologists nor the Marine Station director, Matt Murphy, probably envisaged the continuation of the survey for at least 28 years. With the benefit of such a large database, the early data have not yet been included in any analysis and these years (1975–1980) have come to be recognised as being effectively a pilot study. For any long-term programme this period is very important, as the appearance of inconsistencies in subsequent data collection can easily skew the results if not mitigated during data handling. However it must be stressed that the high standards maintained on the programme have, as noted above, allowed real biological change to be recorded.

Although the concept of long-term monitoring is a commonly held one, the period of time required to qualify as long term is repeatedly debated. It may be tempting to think of five or six years as sufficient to qualify as "*long term*", but at Sherkin for example, although a five year study period (Leslie, 1986) detected seasonal variation in ephemeral green algae similar to the twenty year findings, it was not long enough to detect the seasonal variation in *Littorina obtusata* revealed by the larger twenty year database. The major differences in species abundance recorded in the 1980s and 1990s highlight the ease with which different interpretations could have been made had the work stopped in 1990 for example. What then is the minimum time frame required for long-term monitoring? Any programme certainly needs to cover more than one cycle of its longest cycling species and by default the longest cycling species needs to be identified at the outset. On Sherkin this would be an 11 year minimum period to account for barnacles (10 years) and flat periwinkles (11 year maximum cycle recorded).

Overall conclusions from this work reinforce the view that although the functional behaviour of species can be used to define the rocky shore environment, each shore is unique in its combination of biological and physical factors. Rocky shores on Sherkin are under stress from some factor or combination of factors, and productivity of the rocky shore community has declined since the late 1980s. Long-term integrated ecosystem monitoring is essential to understand our natural environments. Despite the existence of a 23-year data set resulting from a 29-year rocky shore survey programme, monitoring should continue beyond this. We have not yet answered the question "What is normal?"

***Gillian Bishop** is based in Scotland working as an environmental consultant within the oil industry. This work involves her in environmental assessments and monitoring of the offshore marine environment around oil exploration and production operations; and brings her into contact with many people concerned about the marine environment – including regulators, local councils, fishermen, NGOs and marine scientists. She has been involved in Sherkin and the island's rocky shores since 1975 (and has just finished writing about this long-term monitoring programme).*

3.12
LONG-TERM FISHERIES MONITORING WITH EMPHASIS ON THE STRIPED BASS (*MORONE SAXATILIS*) FROM THE HUDSON RIVER

Byron Young
Kim A. McKown & Julia M. Brischler, Bureau of Marine Resources, New York State Department of Environmental Conservation, 205 North Belle Mead Road, Suite 1, East Seatuket, NY 11733, USA

Introduction

The Atlantic striped bass (*Morone saxatilis*) is one of the more popular inshore commercial and recreational fish species found on the Atlantic coast of the United States. The species ranges from the Saint John's River in Canada to the Saint John River in Florida (Versar Inc, 1990; Collette & Klein-MacPhee, 2002). Major producing areas for this anadromous species are the Chesapeake Bay, the Hudson River, the Delaware River, and the Roanoke River (Figure 1). Non-migratory populations can be found in many southeastern Atlantic coastal rivers. The species has also been introduced into many large freshwater impoundments and the Pacific coast (Versar Inc, 1990).

Beginning in the late 1930s with Daniel Merriman's doctoral dissertation, entitled "Studies of the Striped Bass (*Roccus saxatilis*) of the Atlantic Coast" (Merriman, 1941), and continuing to the present, striped bass have been the focus of a very ambitious set of research and assessment programmes. Beginning in the late 1970s, a series of management meetings and management programmes have lead to a collaborative collection of environmental and management data for this species. Increasing concerns over the health of the Chesapeake Bay striped bass stock lead the Atlantic States Marine

Fisheries Commission (ASMFC) to organise a symposium in 1978 to review the existing knowledge and research efforts relative to the striped bass resource for the Atlantic coast. Over one hundred researchers from the coastal states, federal agencies, universities and private research groups, along with invited members of the commercial and recreational fishing public, met laying the foundation for the first ASMFC Fishery Management Plan (FMP) for striped bass, which was adopted in 1981. The FMP was quickly followed by several amendments to address weaknesses in the management programme, as well as the continued decline of the stock. Subsequent amendments have adjusted the fishery management options to allow for increased protection or harvest consistent with the status of the stock.

As a result of the management programme developed under the auspices of ASMFC there are a growing number of long-term data sets available regarding striped bass coastal migratory stock, including juvenile indices from the major producing areas: the Chesapeake Bay, the Hudson River, and the Delaware River. The Chesapeake Bay young-of-the-year (YOY) striped bass survey contains the longest continuous data set commencing in 1954 (Versar Inc, 1990). Richards *et al.* (1999) described the research and population monitoring efforts directed toward determining the reasons for the decline of the striped bass. These efforts have resulted in a number of juvenile and adult data collections that are currently fifteen to twenty years long. Coincident with the research directed toward striped bass management, there has been an equally impressive body of research directed at assessing environmental and anthropogenic impacts, especially in the Hudson River. This paper was prepared to address the theme of the conference entitled "The Long-Term Monitoring of the Marine Environment", and will focus on striped bass data sets

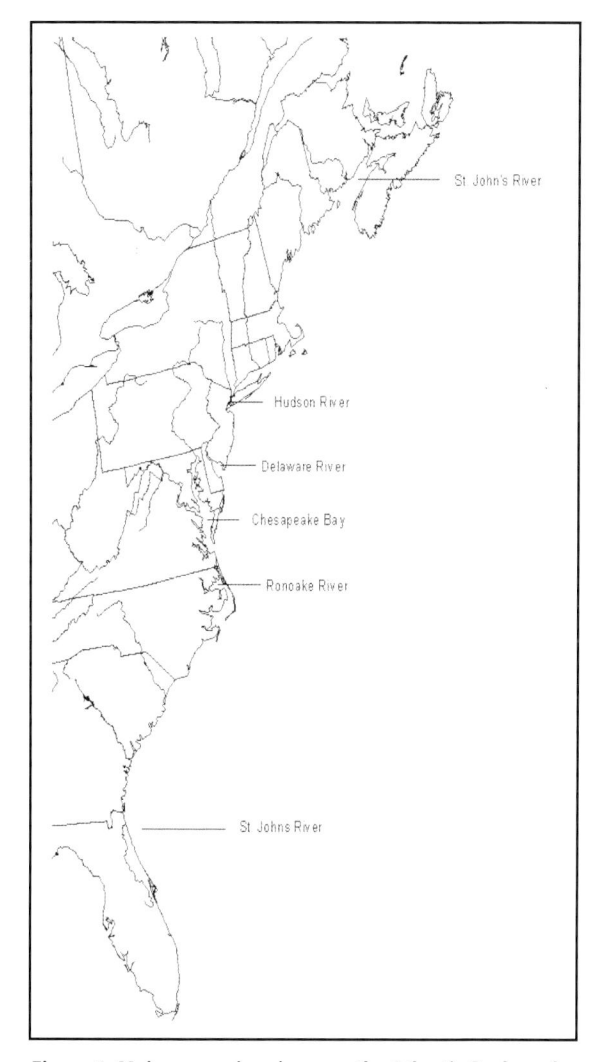

Figure 1: Major spawning rivers on the Atlantic Seaboard.

from the Hudson River estuary. During the past three decades, a large amount of research has been directed toward the Hudson River striped bass in relation to impact assessments and fisheries management. We will focus on this research and monitoring and relate it to coastal fishery interactions.

The Hudson River Estuary

Cooper, Cantelmo and Newton (AFS 1988) presented an overview of the Hudson River and the Hudson River Estuary. The tidal portion of the Hudson River estuary extends 243 km northward from New York City at the Battery, on Manhattan Island, to the Troy dam (Figure 2). Barnthouse *et al.* (1988) describe three major physiographic zones for the Hudson River: (1) a shallow, wide, and brackish lower zone extending from the George Washington Bridge to the upper end of Haverstraw Bay; (2) a deep, narrow, and usually freshwater middle zone from Haverstraw Bay through the Hudson Highlands to Newburgh Bay; and (3) a shallow, wide, freshwater zone from Newburgh Bay to the Troy Dam. Salinity varies with freshwater input, tidal currents and basin morphology. Typically, salinity is lowest during the spring runoff period. The waters off Manhattan are mesohaline (5–18 parts per thousand) in the Spring but increase to polyhaline (18–30 parts per thousand) during the summer and early fall. The salt wedge is generally found in the Hudson Highlands, near West Point although it may move 120 km up river to the Poughkeepsie area in drought years. Temperature varies seasonally, with the lowest temperatures found during January (0.6 to 1.3 °C) and the highest temperatures found during July and August (22 to 29 °C). Higher temperatures may occur in shallow portions of the estuary.

Suszkowski and Waldman (1996) describe the Hudson River basin, containing one of the most heavily populated regions in the nation, as having been subjected to serious pollutant contamination. Sewage discharge and industrial chemicals are believed to

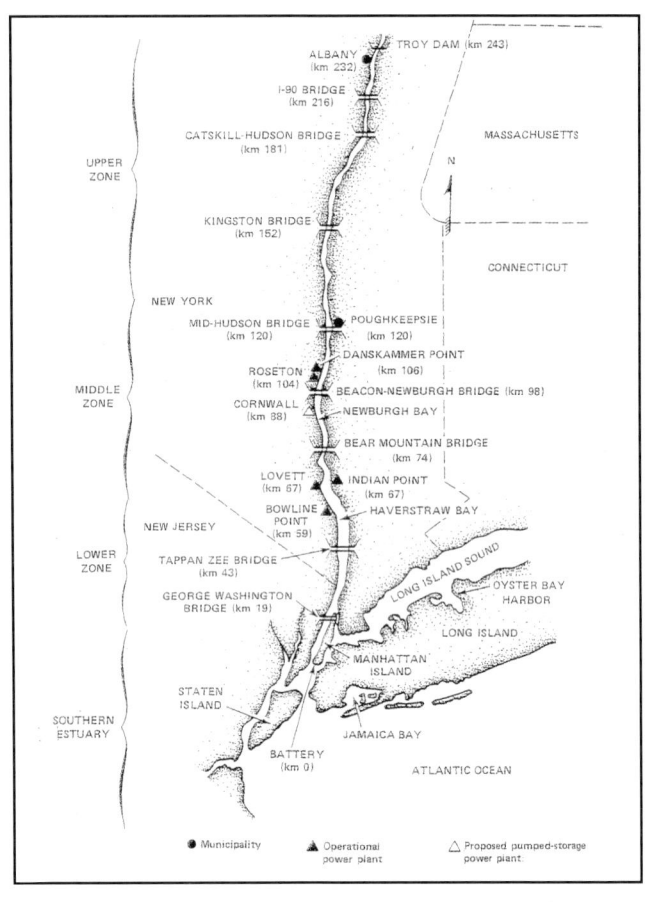

Figure 2: Map of Hudson River Estuary (From Barnthouse, 1999)

have had a profound effect on the ecosystem of the Hudson River during the past century. Mancroni *et al.* (1989) reviewed the recent water quality trends for the mid-Hudson River Estuary. They reported that dissolved-oxygen values reflect an inverse relationship with water temperature. Peak dissolved oxygen occurs in late January/early February when water temperatures are at there lowest, while the lowest dissolved-oxygen values occur during June through October when water temperatures are at their highest. The mid-Hudson River estuary dissolved oxygen levels average around 6 mg/L during the warmer months. Brosnan and O'Shea (1996) describe the long-term improvements in water quality due to sewage abatement in the lower Hudson River. They report a marked improvement in dissolved oxygen from river kilometre 0 to 25. Summer dissolved-oxygen minimum levels have increased from less than 1.5 mg/L in the early 1970s to greater than 3.0 mg/L in the 1990s, and that the duration of any hypoxia during the summer months has been reduced.

Pollutants and nutrient loading has been a major problem for the Hudson River. Major municipal wastewater discharges occurred in the Albany area (River km 240) and in Manhattan (River km 0 to 20). During the mid-1970s municipal wastewater discharge was 4,000,000 cubic metres per day and included some 200 tons of biochemical oxygen demand (BOD) daily, most of which originated in the New York metropolitan area (Cooper *et al.*, 1988). Industrial wastewater discharges have had a major impact on the estuary, primarily the discharges of polychlorinated biphenyls (PCBs) into the upper Hudson River above river kilometre 243. The bulk of the PCB pollution originated near Schenectady on the upper Hudson above the dam at Troy, and has been documented as a contaminant in many Hudson River species, including the striped bass. Sloan *et al.* (1988; 1995) and Sloan & Hattala (1991) described the trends in PCB contamination in striped bass in the Hudson River and the Marine District of New York. Due to PCB contamination, the Hudson River commercial fishery for striped bass was closed in 1976, and remains closed to this day. The recreational fishery is open, but operates under strict human health advisories. Other chemical pollutants have impacted the Hudson River estuary. These, however, have not had the demonstrated impact that PCB's have shown.

The Hudson River Data Sets

Over the past three decades, there have been numerous research and monitoring surveys conducted in and around the Hudson River relative to striped bass. Much of the work addresses concerns relative to the combined affect of seven major power generation stations on the Hudson River. The results of this work are reported in various Annual Year Class Reports prepared by Hudson River utilities and in three compendiums on this research (AFS, 1988; Hudson River Environmental Society, 1988 and 1992). A fourth review of this work was presented at a symposium entitled "Fish Community of the Hudson River" held at the 2002 American Fisheries Society's Annual Meeting held in Baltimore, Maryland. The symposium provided descriptions of the physical and chemical characteristics of the Hudson River estuary, the recent history of the fish community, the status of key species, and some of the interactions among the species that determine the community and population patterns. The proceedings were not available for review during the development of this manuscript. The publications and symposia mentioned above all focus primarily on the efforts of various Hudson River utilities to describe the impacts of power plants on the Hudson River estuary. Barnthouse et al. (1988) described the work conducted on the Hudson River relative to Power Plant impact as "one of the most ambitious environmental assessments ever performed." AFS Monograph 4 (1988) describes more than 15 years of studies associated with the Hudson River Power Plant issue. The Hudson River Power Plant data set includes over thirty years of scientific research. This information is reported in various Year Class Reports, the Draft Environmental Impact Assessment for State Pollutant Discharge Elimination System Permits for Bowline Point, Indian Point 2& 3, and Roseton Steam Electric Generating stations (Central Hudson Gas & Electric Corp. et al., 1993) as well as the Final Environmental Impact Statement for the same power generating stations New York State Department of Environmental Conservation (NYSDEC, 2003).

During the same time period, the NYSDEC conducted research work in the Hudson River estuary. This work focused on developing data relative to the management of striped bass in New York and along the Atlantic coast. The primary repository for this information resides with the Atlantic States Marine Fisheries Commission's (ASMFC) striped bass fishery management programme. The NYSDEC has conducted a series of striped bass monitoring programmes since 1973. These have been funded under grants from the National Marine Fisheries Service and the US Fish and Wildlife Service. The Hudson River time series began in 1973 as a juvenile striped bass tagging programme (Young, 1976), which was modified to form the basis for the current Hudson River young-of-the-year index of abundance in 1976 (Young, 1979). The programme was modified to its current standard format in 1980 (Young, 1980). It was expanded temporally in 1985 and has run continuously since that time (McKown, 2001). Collections were made using a 61 metre x 3 metre seine with 6 mm square mesh in the wings and 5 mm square mesh in the pocket. A second long-term juvenile striped bass data set (McKown & Brischler, 2002) was commenced in 1984 with a seine survey for yearling and older striped bass in the waters around western Long Island (Young, 1984). Collections were made using a 61 x 3-metre seine described above. The two time series were used to validate the Hudson River young-of-the-year striped bass index (McKown, 1991), and they are two of several juvenile indices that are used for determining the status of the striped bass stock. The NYSDEC also conducted a juvenile trawl survey from 1981 to 1990 in the lower Hudson River Estuary. Trawl collections were made at standard stations on offshore shoals (1.8 to 9 metres in depth) with a 7.9 metre head rope, Carolina wing bottom trawl. Stretch meshes in the body, cod end and cod end liner were, 3.8 mm, 3.2 mm, and 1.3 mm. Tows were 5 minutes against prevailing current. Mean vessel speed was 1.2 m/sec (Kahnle, 1988).

The NYSDEC has also conducted an adult stock status survey since 1981. This survey has two components, a Spring adult spawning stock survey (Kahnle, Stang, Hattala & Mason, 1988), and a fishery dependent survey examining the by-catch of striped bass in the American shad gill-net fishery (Kahnle, Hattala, & Stang, 1988). The spring spawning stock survey is conducted using a 152 x 3.7 m seine with 68 mm square mesh wings and a 25 mm square mesh bag. The department has also use a 31 x 3 m small-mesh seine to monitor the production of American shad (*Alosa sapidissima*) in the mid and upper Hudson River (Kahnle, Hattala & Stang, 1988). This effort has run since 1981.

Heimbuch *et al.* (1988) evaluated six indices of Year Class strength of Striped Bass in the Hudson River. These six indices included three Hudson River utility surveys; An ichthyoplankton survey using a fixed-frame trawl with 505μm mesh; and two juvenile surveys: one using a 31 x 3 m beach seine survey with 1 cm square mesh and a 0.5 cm mesh pocket; and the second the Fall shoals survey, which has used a variety of towed gear: a 1 m tucker trawl to sample the shoals and 3 m x 3 m mesh beam trawl with 3.8cm mesh in the body, 3.2 cm mesh in the cod end and 1.3 cm mesh in the cod end liner prior to 1985. After 1985 the survey has been conducted with an epibenthic sled. It also included three agency surveys, a 200-foot beach seine survey, a 100-foot beach seine survey, and a bottom trawl survey. Each programme produced an estimate of abundance or relative abundance for either larval or juvenile Hudson River striped Bass stock. The usefulness of each survey is dependent upon the question being addressed by the survey. The utility surveys assess the impact of power generation at the population level. The NYSDEC surveys provided an annual index of relative abundance for juvenile striped bass to examine trends in recruitment over time. These data are used along with other coast wide monitoring data to assess the status of the coast wide striped bass stock. The current assessment integrates these data sets in a virtual population analysis (Atlantic States Marine Fisheries Commission, 2002).

Heimbuch *et al.* (1989) concluded that CPUE estimates from beach seine or offshore sampling programmes alone do not consistently represent young-of-the-year abundance. McKown (1991) found that a combined seine and trawl index correlated best with yearling abundance. Despite this, the surveys continue to be useful for research and assessment of striped bass and other species. Students at a number of colleges and institutions to examine species interactions and relationships currently use the information collected during these monitoring programmes.

The survey work conducted by the NYSDEC in the Hudson River (McKown, 2001) and around western Long Island (McKown & Brischler, 2002) has demonstrated an expansion of the range of young-of-the-year striped bass. Prior to the late 1980s, juvenile striped bass (YOY) were only observed in and around western Long Island during years of high relative abundance. However, following the implementation of stricter management during the mid-1980s, YOY striped bass have been observed in that area on an annual basis regardless of the relative strength of the Hudson River YOY indices observed for the Hudson River estuary (Suarez, 2003).

Discussion

The various data described above have been used by the Hudson River Utility companies and their consultants to assess Power Plant impacts, including estimates of juvenile impacts through Conditional Entrainment and Impingement mortality rates for each of the major power plants, and estimates of subsequent adult impacts through population models. Likewise, the agency data in conjunction with the utility data have been utilised in coast-wide assessment of the striped bass migratory stock, for Hudson River stock specific assessments, and to develop management targets. The whole data set has been used to examine ecological questions such as; predator/prey relationships (Buckel et al., 1999) and competition (Buckel & McKown, 2002).

The data referred to in this document span more than three decades of research and involves thousands of individual records. These records include information down to the individual station catch information. The information referred to in this document relate to striped bass, but information exists on all species encountered within the Hudson River estuary and forms an exceptional set of fisheries, physical, and chemical data for the estuary. Most of the information is available in various reports and is presented in various ways to address a particular question or series of questions. The agency data are public domain information and is available in individual record format. The utility data have recently entered the public domain, though remain difficult to access. Aggregated utility information is available in numerous reports prepared to meet annual permitting requirements. There is a need, however, to see all of the utility data reside in the public domain and to have annual updates available in a timely manner. Current researchers are beginning to examine long-term data series using newer more sophisticated techniques. A symposium at the 2003 American Fisheries Society meeting in Quebec City provided some insight into the future of this effort. Presentations focused on standardisation, calibration of research gear, multispecies-multigear approaches to monitoring, and the value of measurements beyond your target species, and survey design and assessment.

There is a series of Estuary Management plans that encourage and support long-term monitoring and multi-partner studies including the Long Island Sound Study, the Peconic Estuary Program, the New York/New Jersey Harbor Estuary Program, and the Hudson River Estuary Management Program. During 1987 the New York State legislation passed the Hudson River Estuary Management Act, which directs the New York State Department of Environmental Conservation to develop a management programme for the Hudson River estuarine district and associated shore lands. One of the Estuary Program objectives is to develop a coordinated long-term monitoring programme for the estuary. The wealth of information currently available, and the apparent willingness of researchers to share their information, provides a strong foundation for achieving the programme's goals and objectives.

It is encouraging to see the growing interest in maintaining long-term environmental data. The data set referred to above, and the data sets available around the world, have become our guide to understanding the future by viewing the past information. The improvements in standardisation, more timely reporting of results through electronic data entry, improved coordination of data collection, and the utilisation of multi-species and ecosystem models and assessments make these data sets more valuable to future generations. A couple of examples that support this view include the efforts of the ASMFC in developing a coordinated inshore trawl survey. The Southeast Area Marine Assessment Program (SEAMAP) has been conducting a collaborative trawl survey throughout North Carolina to Georgia inshore ocean waters for several years. Currently the states from North Carolina to Maine have been working to develop a Northeast Area Marine Assessment Program (NEAMAP). These two programmes are attempting to reduce redundancy of effort by creating a programme that relies on one or two research vessels to conduct a regional inshore survey of the marine environment.

In conclusion, it is very exciting to see the growing body of research that is employing new approaches, or retrospective analysis with long-term data. The 2003 American Fisheries Society (AFS) Annual meeting, held in Quebec City, hosted a symposium entitled "Large-scale fisheries independent surveys: Looking to the future by learning from the past." The topics addressed were standardisation,

- gear efficiency,
- lessons learned from conducting long-term monitoring,
- problems associated with initiating and maintaining surveys,
- survey design,
- data utility,
- changes in sampling protocol,
- gear comparison and
- compatibility and timeliness of data acquisition.

A recent paper by Latour *et al.* (2003) reviewed four multi-species modelling approaches and their potential as fisheries management tools. An important observation in this paper is that, *"model results need to be compared to independent time series data (e.g., biomass, abundance) not used in model calibration. The model must reasonably track the data for one to have confidence in its function. Thus, these types of time series data should be considered as a data requirement in addition to those used to parameterise the model."* The growing number of long-term fisheries data, environmental data, and organisational goals centred on the collection of new long-term monitoring of the environment laid the foundation of data that will aid researchers for years to come. We look at this as the sunrise of a new era in fisheries research.

Byron Young *is a marine biologist working with the New York State Department of Environmental Conservation.*
As Chief of the agency's Finfish and Crustaceans Unit he is responsible for program oversight covering a variety of marine, anadromous and crustacean fisheries research and management projects. He has been involved in review of the Hudson River Power Plant impact assessment. He is responsible for project oversight of the agencies three striped bass projects; Project include the Hudson River striped bass Young-of-the-year survey, the Eastern Long Island striped bass tagging study, and the western Long Island juvenile striped bass survey. He has represented New York on the development of a draft recovery plan for Shortnose sturgeon, the Striped Bass Fishery Management Plan and the American shad and River Herring Fishery Management Plan.

3.13

LONG-TERM MONITORING OF MARINE PHYTOPLANKTON AT SHERKIN ISLAND MARINE STATION

**Geraldine Reid
Curator of Diatoms,
Department of Botany,
The Natural History Museum,
Cromwell Road, London,
SW7 5BD, UK**

Marine phytoplankton are microscopic 'plants' that are found floating in the sea and form the base of the food chain, in that most organisms depend on them, or something that has fed on them, for sustenance. They have practically no independent means of movement in the sea and as a consequence are carried around by local currents. Long-term monitoring of marine phytoplankton is rare and mainly restricted to a few stations along the coastline. In most of these monitoring programmes only chlorophyll measurements are available with any regularity. This will only give you an idea of the biomass in the water and not of the actual biodiversity. Chlorophyll measurements do not give any information about what the population is made up of or of the proportional contribution of each species to the overall population density.

One of the few long-term programmes is The Continuous Plankton Recorder, which started in 1931 and covers the whole of the North Sea and North Atlantic (Colebrook, 1960; Hardy, 1935). Due to the chosen mesh size of the retaining silk (270 µm), quantification of phytoplankton species is restricted to a few very large species of diatom and dinoflagellate, and it is not possible to determine distributions and population trends for the many thousands of other species.

The phytoplankton-monitoring programme at Sherkin Island has been in operation since 1978. The programme has changed little over the years. Water samples are collected using a Nansen bottle, which is passed through the water column and collects discrete water samples at various depths (0.5, 2.5, 5, 10, 15, 20, 25, 30 and 50 metres, depending of course on the maximum depth of the site). The samples are fixed with Lugols iodine, a gentle preservative that tends not to distort the cells, unlike many other preservatives. The samples are then taken back to the lab and set up in sedimentation chambers to count the numbers of cells per litre found at the various depths. The phytoplankton net, with a mesh size of 25 µm, is pulled up through the water column to give the overall composition of the plankton throughout the sampled water column. This sample is taken in case any species occupy a discrete layer that would be missed by taking samples only at defined depths. The net samples are not initially preserved and are taken straight back to the lab for identification whilst still alive and then subsequently fixed. All samples are permanently archived at the Sherkin Island Marine Station.

The sites consist of nine 'Bay' stations around Roaringwater Bay and four 'South' stations which define a transect running immediately south of Sherkin Island (Figure 1). Station 1 is located at the Gascanane Sound, one mile south of Sherkin, and Stations 2, 3 and 4 are located at 4-mile intervals (4, 8 and 12 miles south of Sherkin); all stations are sampled every 10–12 days. The frequency of sampling at Bay 8, the landing stage, was increased to every 4 days in 1996 in order to more finely resolve the dynamics of the ever-changing plankton. This site is easier to access than the other sites so it is possible to sample here even in quite rough weather.

The monitoring at Sherkin Island Marine Station started in the early years because of a large persistent bloom of the dinoflagellate *Karenia mikimotoi* (Miyake & Kominami ex Oda) Hansen & Moestrup[1]. This resulted in a strong discolouration in the water due to the high cell densities in the water column, a phenomenon which is often referred to as "red tides" and which can cause massive fatalities of benthic fauna and fisheries (Leahy, 1980; Parker, 1980).

The Sherkin Island Marine Station's programme had the foresight to begin monitoring the whole of the plankton rather than just this one species. Due to constraints of time and money, many monitoring programmes only enumerate the problem taxa and record perhaps at best a few other dominant species. Over time, various species may become problematic and if one has only limited species lists it may not be possible to describe or understand the changing ecology.

By looking at the time-series data one can see that the numbers of *K. mikimotoi* have drastically decreased over the years (Table 1). For example, in 1978, cell densities were over 7 million cells per litre

Table 1: Maximum number of cells

Year	No. Cells per litre
1978	7 700 000
1979	5 708 000
1980	10 900
1984	12 700
1988	160 200
1992	18 000
1996	650 000
2000	4 919 400
2003	300

---

[1]This species was previously referred to as *Gyrodinium aureolum* Hulburt and *Gymnodinium mikimotoi* Miyake & Kominami ex Oda. Advances in taxonomy have lead to this change in the generic and specific name.

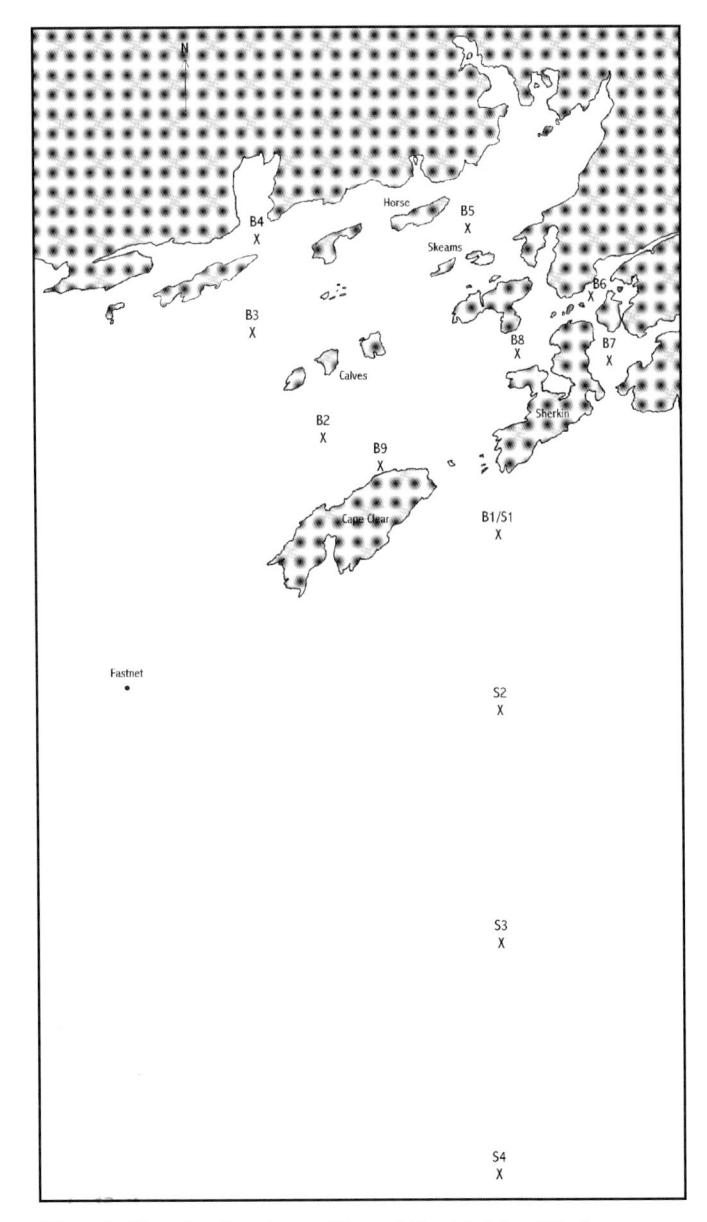

Figure 1: Map showing the positions of Sherkin Island Marine Stations phytoplankton monitoring sites.
B1–B9 = Bay sites; S1–S4 = South sites

(Roden *et al.*, 1980) and in 1979 were over 5 million cells per litre (Roden *et al.*, 1981). Numbers over the next 20 years were considerably reduced (Table 1). High cell densities recurred in 2000, with cell numbers again exceeding 4 million. This bloom was less widespread and persistent than the earlier blooms.

However, as the economic and ecological nuisance caused by *K. mikimotoi* has declined, other problem organisms have taken its place. There has been an increase in the problems caused by the *Dinophysis* group of dinoflagellates. This group is responsible for causing diarrheic shellfish poisoning (DSP). They produce Dinophysistoxin-1 (DTX1) and okadaic acid. Shellfish feeding on these dinoflagellates accumulate the toxins in their flesh, which when consumed by humans or animals causes severe diarrhoea, nausea, vomiting and abdominal pain after approximately 30 minutes.

The economic effects on the shellfish industry have highlighted the problems caused by these species, resulting in the closure for harvesting in the past, of some shellfish farms for up to 10 months of the year. The southwest coast of Ireland is an important site for both finfish and shellfish aquaculture. Within Roaringwater Bay there are several shellfish farms which yield approximately 1000 tonnes of mussels and 100 tonnes of oysters annually.

The importance of time and space scales

This paper highlights the importance of considering the larger time-scale rather than just a snapshot in time, which is the usual proviso of phytoplankton monitoring. I illustrate the importance of sampling at various depths within the water column and using multiple stations rather than just one fixed point on a large shoreline. *Dinophysis acuta* Ehrenberg is used as a working example, and its distribution at three of Sherkin Island's sites is considered: at Bay 2, Bay 8 and South 2 (Figure 1). To put the figures in context, it should be noted that whereas vast numbers of *K. mikimotoi* in the millions of cells per litre caused earlier problems (Table 1), as little as 100 *D. acuta* cells per litre can have a problematic affect.

Figure 2 shows the number of *D. acuta* cells per litre plotted against the date from 1980 through to 2003 at all sampling depths at Bay Station 2. No one depth predominates or maintains the maximum cell numbers consistently across all the years of sampling (Figure 2). The maximum peak for the 25 years, of 4900 cells per litre at 15 m depth, occurred in 2002.

Figure 3 gives the cell densities of *D. acuta* for Station South 2 from 1980 to 2003. Here, the peaks may appear less frequent than at Bay Station 2 (Figure 2) but the scale on the axis in Figure 3 is considerably larger, with a peak of 17,600 cells per litre at 25 m on 26 August, 2002.

A comparison of Figures 2 and 3, apart from the major peak in Figure 2 of 4,900 cells per litre, shows that all of the peaks were below 2,300 cells per litre, with most being under 1000 cells per litre. The magnitude of the cell numbers in Figure 3 is far greater. Blooms of *D. acuta* appear to develop offshore and then move into the Bay. However, blooms develop in the Bay, as well, but at a smaller scale. The blooms are developing at depth rather than in the surface. The magnitude of cell densities at South 2 (Figure 3) is far greater than at Bay Station 2 (Figure 2).

Figure 4 shows only the 2002 data for *D. acuta* at South Station 2 and for clarity plots only two depths: 25 m, where we obtained the maximum cell densities, and 2.5 m. It is evident from Figure 4 that blooms would be missed should sampling be limited to just the upper water column. Whilst the bloom of *D. acuta* was occurring at 25 m, there were no cells present at 2.5 m. A consideration of other years in the Marine Station's

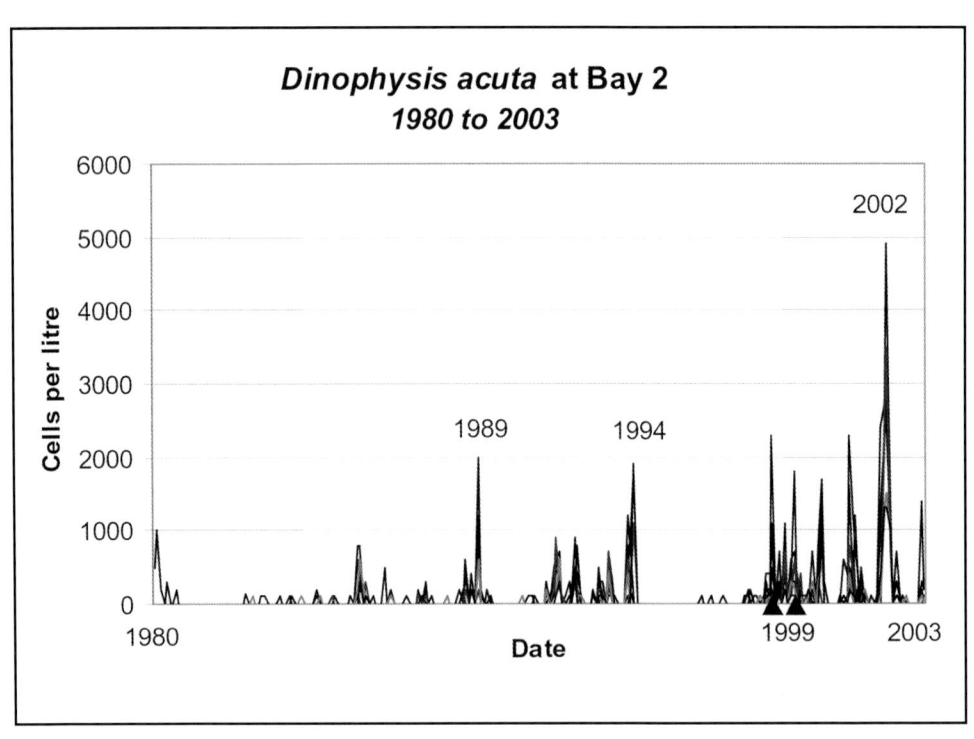

Figure 2: Cell densities of *Dinophysis acuta* at Bay Station 2 from 1980 to 2003.

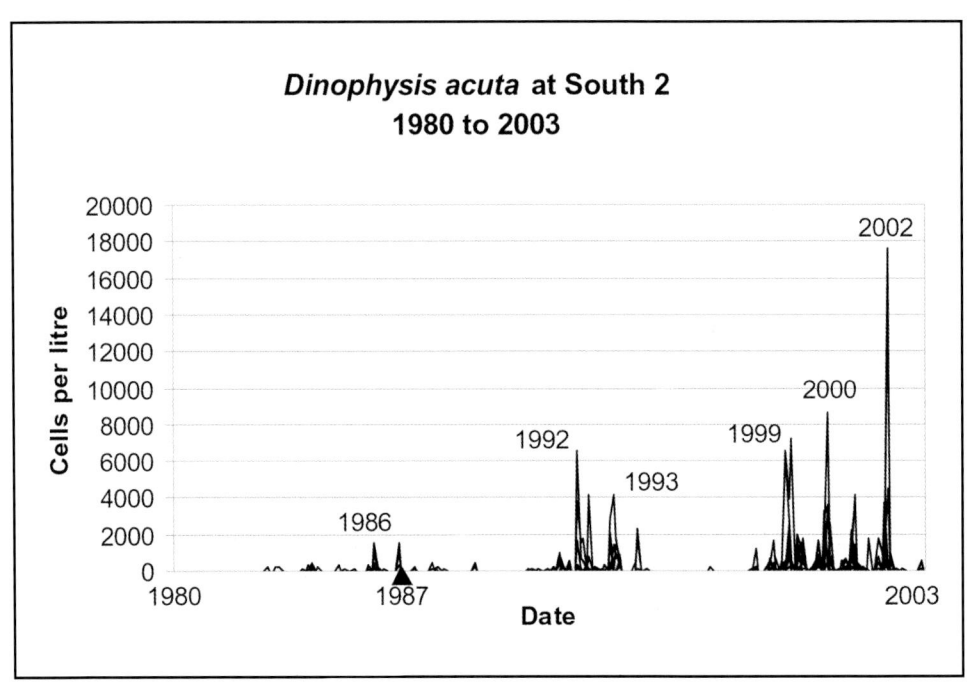

Figure 3: Cell densities of *Dinophysis acuta* at South Station 2 from 1980 to 2003.

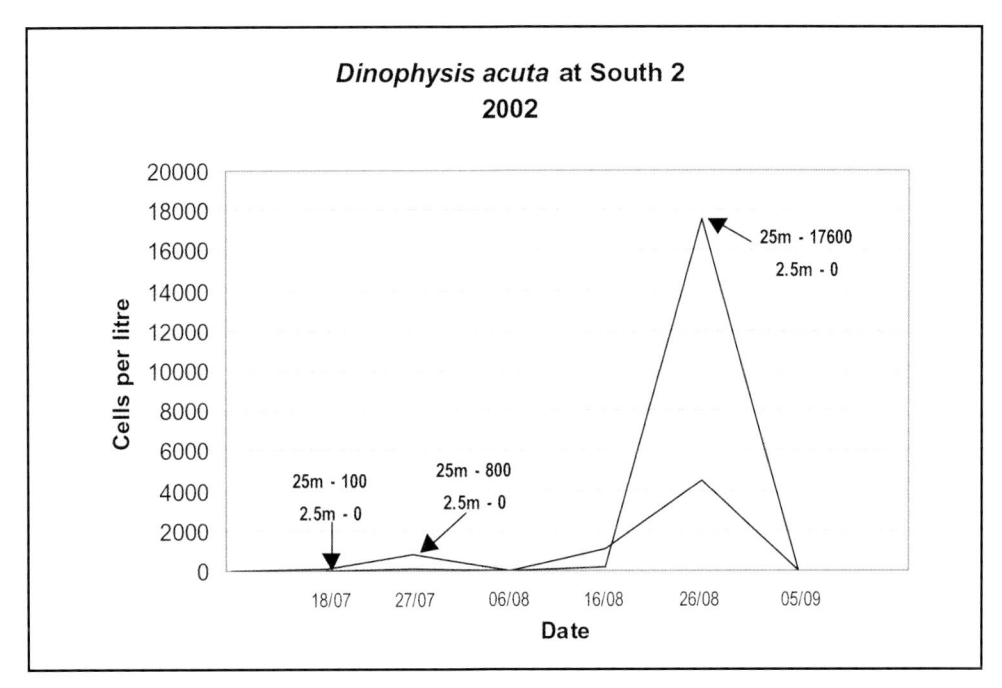

Figure 4: Cell densities of *Dinophysis acuta* at South Station 2 for 2002 at depths 2.5m and 25m.

data shows again that it is important to sample at various depths in order to capture all the population dynamic information.

Bay 8, the landing stage situated just off the marine station (Figure 1), has been sampled every four days since 1996. This has increased our knowledge of the population dynamics by giving greater resolution of species variation over time in the Bay (Figure 5).

Figure 6 shows data for *D. acuta* at the landing stage in the Bay only in 2002, to highlight in more detail what happened to the population that year. Arrows along the bottom indicate the normal sampling dates of the Bay trips with the other points being due to the increased sampling (every 4 days) at the landing stage. Figure 6 illustrates the increased knowledge gained from regular sampling. It can be seen that a lot of the interesting increases took place when the regular trips were not taking place (Figure 6). Having a site that is frequently monitored can add a great deal of information that would otherwise be lost due to the constraints of time and money.

Taxonomy

Careful, accurate, and consistent identification of the taxa is critical for the reliability of this work. Within any monitoring programme it is vital that everyone within the group is calling the organism by the same name, otherwise no meaningful information can be gleaned about the organisms of interest. It is very important to have a taxonomic expert available to those working on ecological surveys.

This concept may be highlighted using *D. acuta* as in the above examples. Figure 7

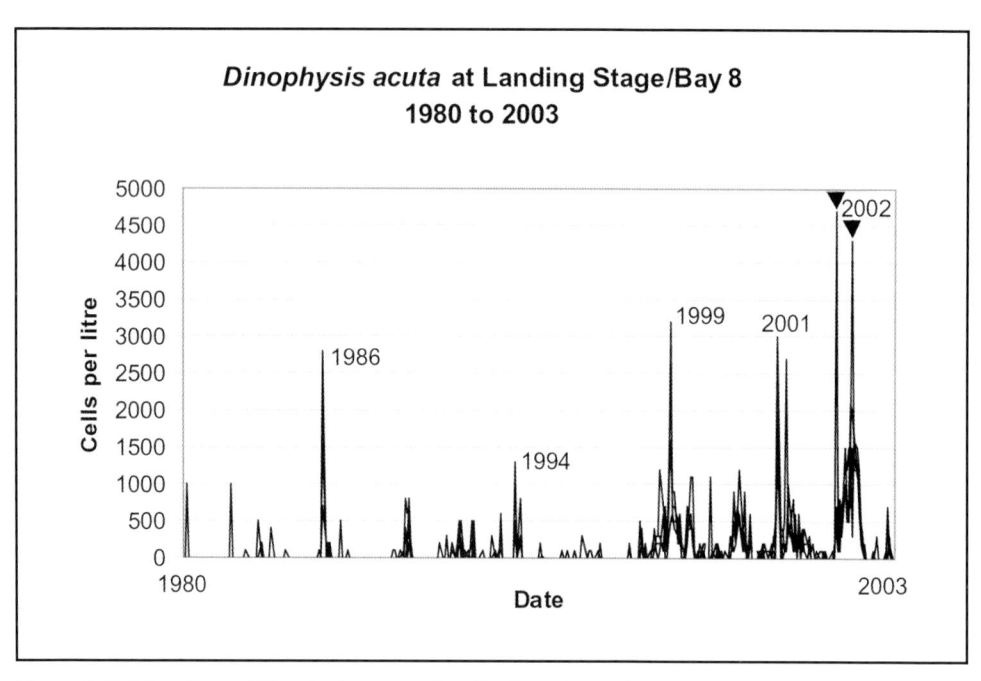

Figure 5: Cell densities of *Dinophysis acuta* at Bay Station 8 from 1980 to 2003.

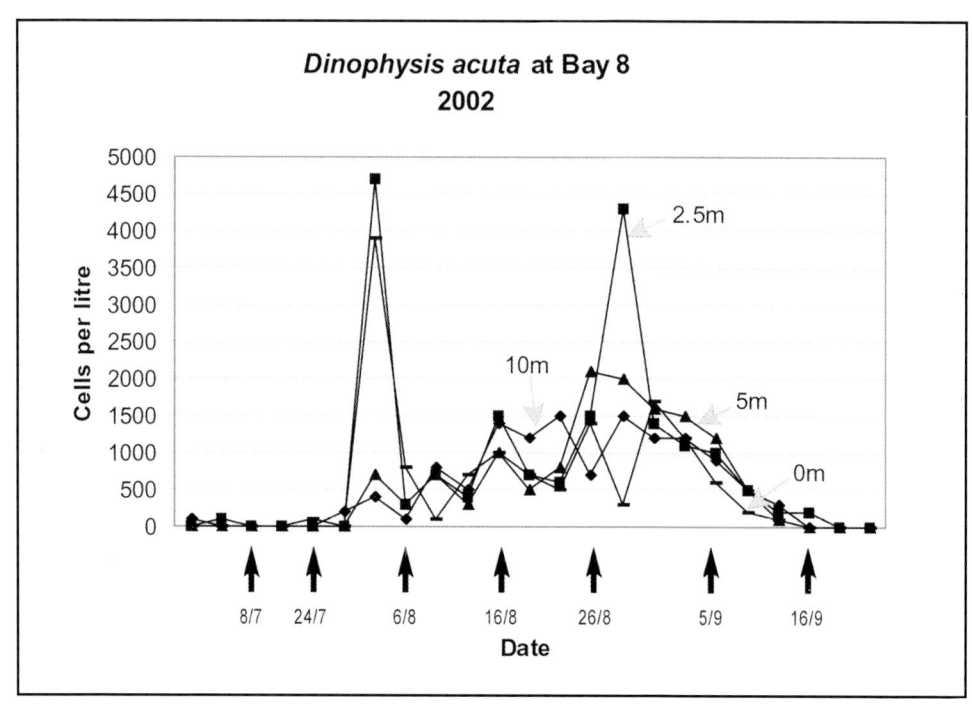

Figure 6: Cell densities of *Dinophysis acuta* at Bay Station 8 for 2002. Arrows along the bottom of the graph indicate the dates of the routine Bay trips (8, 24 July, 6, 16, 26 August, 5, 16 September 2002)

illustrates *D. acuta* (Figure 7A) and *D. norvegica* (Figure 7B). Superficially they look quite similar, with heavily areolated cells and strongly developed cingular lists (c) (Figure 7), but on careful inspection they are very different. In *D. acuta* the left sulcal list (LSL) joins the cell below the widest portion of the cell, whereas in *D. norvegica* it joins the cell at the widest part. The ventral margin of the cell in *D. norvegica* is concave (as indicated by the arrow, Figure 7B). But, more importantly than these subtle differences in their appearance, their effect on toxicity in shellfish is quite different.

With *D. acuta*, a very small number of cells per litre, in the order of 100–300, can cause problems in shellfish, whereas *D. norvegica* requires considerably more, on the order of 2000.

Within any monitoring programme there should be a call for the inclusion of illustrations and specimens linked to permanently stored samples. When scanning through the literature it becomes very clear that a vast array of ecological papers have been written on the population dynamics of toxic species and other planktonic taxa but very few papers indicate whether they have conserved specimens in stored samples. This is vital, and should be made a priority, so that future generations may check the identifications as advances are made in taxonomy and ideas about the effects of the organism are advanced. As in the case with *K. mikimotoi*, I was able to check the Sherkin archives to find out where we really dealing with *K. mikimotoi* or *G. aureolum* around the Roaringwater Bay by looking at the placement of the nucleus in the cells.

We need to maintain records of all species present in the water column, not just the dominant species. As one can see in the case of *Dinophysis*, problems can be caused

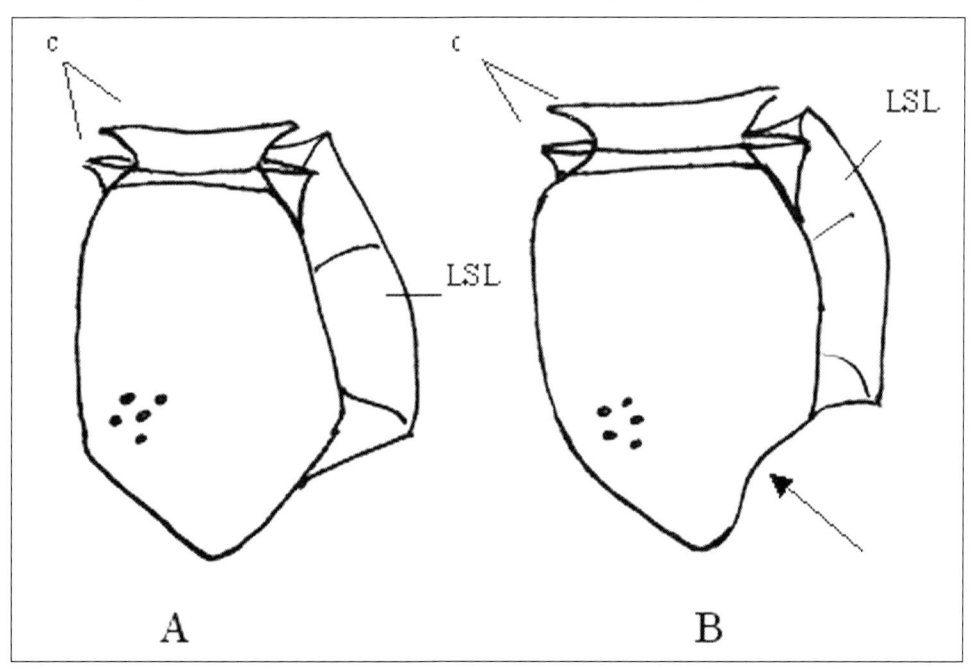

Figure 7: Schematic diagrams of *Dinophysis*. A: *D. acuta*; B: *D. norvegica*.
C = cingular lists, LSL = left sulcal list; arrow indicates concave ventral margin.

even by low cell numbers. In a limited monitoring programme, there would not have been any record of their history until they were linked to toxicity.

The full 25 year data set is currently being databased which will allow us to look at it in an easily accessible way and allow detection of any effects of climatic change on the distribution of phytoplankton. It will enable us to look back at the past and to use the historical perspective as a tool for predicting ecological change in the future.

***Geraldine Reid** is the Curator of Diatoms at the Natural History Museum, London. Her interest in marine phytoplankton started in 1992 whilst working at Sherkin Island Marine Station on the marine phytoplankton of the area. She has returned annually to harmonise and supervise the sampling and identifications of Sherkin Island Marine Station's monitoring programme.*

3.14
LONG-TERM OBSERVATIONS: CRUSTACEANS AND MOLLUSCS IN ATLANTIC CANADA

René Lavoie
Manager, Invertebrate Fisheries Division, Science Branch, Fisheries & Oceans Canada, Bedford Institute of Oceanography, PO Box 1006, Dartmouth, Nova Scotia, B2Y 4A2, Canada

Introduction

The observation of environmental factors by mankind undoubtedly dates back to the beginnings of our species on Earth. Indeed, such observations were constantly made and analysed in the mind of primitive peoples; the resulting knowledge was transmitted through the generations via oral tradition. The right knowledge about air, water, earth and fire coupled with the knowledge of animal life and plant medicinal properties allowed certain tribes to survive, prosper and expand. Others, who did not have the right knowledge withered and disappeared.

This orally transmitted knowledge is rooted in thousands of years of human experience; it is broadly referred as Traditional Ecological Knowledge (TEK). Some practitioners of modern science feel that TEK may have value and they struggle to find a bridge which would join the knowledge chain of TEK and what we call Modern Science.

Modern quantitative science as we know it is relatively young. Long-term data sets combined with an unprecedented analytical power helped by modern computers have brought the capacity to quantify the evolution of the world around us. Too much reliance on these tools can however be dangerous. Models based on past observations can only make accurate predictions if the future is a repetition of the past. Constant and attentive observations of ecosystems as they evolve around us and frequent re-interpretations are necessary to get the knowledge required for adaptive living.

This paper presents some examples which illustrate the value of long-term observations as they apply to some segments of the Scotia Fundy Fisheries Management Area on the East Coast of Canada and offers considerations for the future.

Human interest

The wild harvest fishery is extremely important to the economic base of many communities in the Province of Nova Scotia, contributing more than C$ 1.2 Billion to the provincial Gross Domestic Product. In the year 2,000, the fishery ranked the number one exporter, being responsible for a quarter of the value of exported goods. The fishery is conducted mostly by small companies and independent operators based in the rural areas; as such, the industry provides much-needed employment where few other industries exist. Fishing and fish processing together create approximately 31,000 jobs.

Fisheries in evolution

From 1990 to the year 2000, the landed value of the fishery in the Scotia-Fundy Region (Figure 1) increased from 438 to 618 M Canadian dollars. However, this apparently healthy picture conceals changes which brought significant social and economic consequences to fishermen and to those who depend on their catches to make a living.

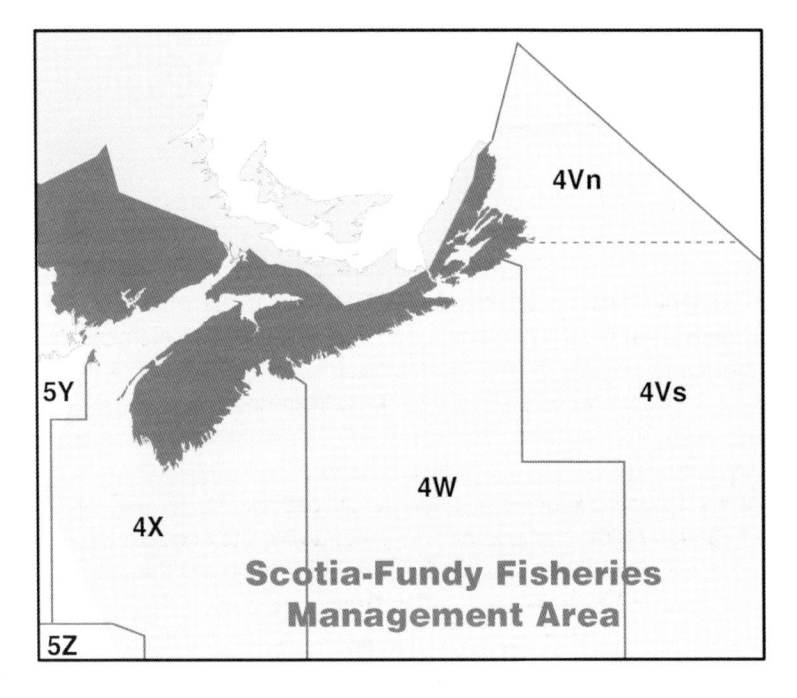

Figure 1: The Scotia Fundy Region of Atlantic Canada

In 1990, pelagics and groundfish represented 47% of the landed value; in 2000, fish represented only 19% of the total, a most dramatic decrease which carried with it severe social and economic impacts. Fisheries comparison pie charts 1990–2000 (Figure 2) show an important shift from groundfish to invertebrate fisheries.

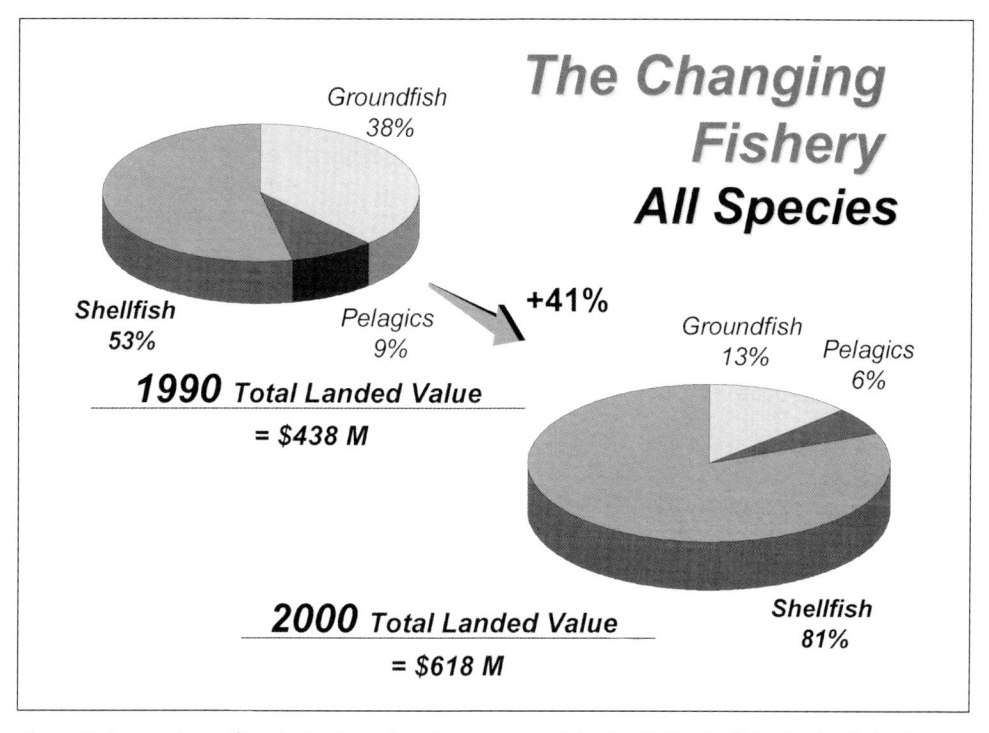

The Changing Fishery
All Species

1990 Total Landed Value = $438 M
- Groundfish 38%
- Shellfish 53%
- Pelagics 9%

+41%

2000 Total Landed Value = $618 M
- Groundfish 13%
- Pelagics 6%
- Shellfish 81%

Figure 2: Comparison of landed values of major segment of the Scotia Fundy fisheries 1990-2000

A search for the causes of the groundfish decline led to the general conclusion that gross overfishing had occurred in most management units. It was difficult to draw conclusions on the relative roles of fishing and the environment on groundfish changes because of variable data quality on the biological features of interest (Angel et al., 1994). The analysis of the data at hand also revealed several problems in the timeliness, completeness and accuracy of the landing and effort data (Etter, 1996). Hence, the obvious need for very special attention to the collection and preservation of complete and accurate data over considerable time if correct interpretation of population trends and adaptive human behaviour are to be achieved.

In the years which followed the dramatic decline in groundfish observed in 1992–1993, the search for explanation was expanded to include an interest in the health of the entire ecosystem. Data on sixty four variables collected through various studies and sources connected with the groundfish fishery were assembled and analysed by a working group of scientific staff led by Ken Frank of the Bedford Institute of Oceanography (DFO, 2003). The study focused on the Eastern Scotian Shelf ecosystem and most data series extended back to at least 1970. Based on the analysis of these extensive data sets, the group was able to reach a number of important conclusions, namely that :

- A major cooling event of the bottom waters occurred in the mid-1980s and persisted for a decade;
- Major structural changes have occurred in the fish community; groundfish have declined while small pelagic species and commercially exploited invertebrates have increased;
- Reduction in average body size of groundfish have occurred and there are currently very few large fish in the population of many species;
- The abundance of grey seals has risen steadily during the last four decades;
- The fishery is increasingly targeting species at lower level in the food web.

Noteworthy as these conclusions may be, they do not provide a complete picture of the ecosystem. They point to significant changes but cannot pinpoint causes. The group realised that, to reach conclusions about an ecosystem based on an evaluation of such a broad scope, more and better data were necessary:

- There are too few reported data on contaminants in water, sediment or biota;
- Knowledge about the diversity of species becomes increasingly sparse at the lower trophic levels;
- There is a need for monitoring benthic invertebrate species.

On a broader scope still, recent authors of fisheries Management Strategy Evaluation (MSE) models acknowledge a strong need for monitoring observations to update management related parameters, for management decision-making and implementation of decisions (Sainsbury *et al.*, 2000; Sainsbury & Sumaila, 2003).

In the future and on a world scale, monitoring environmental factors will be even more crucial because technology has dramatically expanded the outer boundaries of systems. This is now obvious in the areas of fisheries, the ozone layer, and possibly the capacity of the atmosphere to absorb carbon dioxide. In the opinion of some forward looking business writers the global economy had already exceeded carrying capacity of the earth ten years ago (Hawken, 1993).

Modern technology and monitoring lower trophic levels

In recent times, modern technology has expanded human capacity for scale and timing of monitoring environmental factors and data collection. From continuous *in situ* recording of environmental factors like temperature, salinity, wind and currents to remote-sensing of sea surface temperature and chlorophyll-A, voluminous streams of data are now stored into ever expanding computer memories, making large scale studies possible. For example, it has been demonstrated that quantitative data from video recordings can clearly detect major organic enrichment of the benthic environment around salmon farms (Crawford, Mitchell & Macleod, 2001).

The quantification of a link between primary productivity and factors important to fisheries is another example. Remote-sensing, satellite data combined with long-term data sets of haddock (*Melanogrammus aeglefinus*) recruitment of the continental shelf of Nova Scotia shows that larval fish survival depends on the timing of local spring phytoplankton bloom (Platt *et al.*, 2003). These authors noted that two exceptional haddock year-classes (1971 and 1999) coincided with unusually early spring blooms observed through satellites.

Phytoplankton blooms variation can also have dramatic effect on animal groups which spend their entire life closer to the lower levels of the food chain like molluscan bivalve filter feeders. The sea scallop (*Placopecten magellanicus*) provides a good example. The average meat-yield of offshore sea scallops on Georges Bank increased dramatically in 1999 (Figure 3), the same year which produced the latest strong haddock year-class.

These observations and conclusions are very valuable in their own right. They may allow scientists to follow the evolution of ecosystems which support important fisheries as technology increases fishing efficiency and climate changes take their toll. It ought to be noted that once again, data sets collected over long period of time are the raw material from which these studies can be made.

Monitoring the top of the food chain

Ecosystem studies with an emphasis on species near the top of the food chain require even more high quality and consistent data sets to achieve the explanation power required to understand fisheries which are important to humans and the ecosystems which support them.

The American lobster, (*Homarus americanus*) provides a good example. Landing records date back to over a century and show progressive decreases followed by

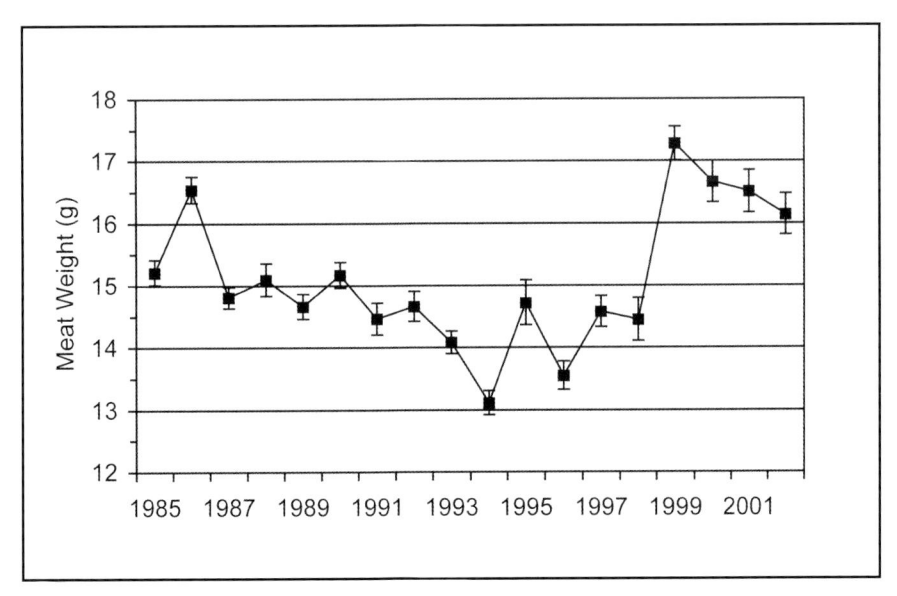

Figure 3: Georges Bank Scallop August Survey Meat Weight Per Standard Shell Height

recoveries which had and still have important consequences for human coastal communities (Figure 4). The lobster fishery is the most important fishery in the Maritime Provinces of Canada at this time; yet, we do not have the data required to understand why landings have come back to historical highs and what to do to maintain them there. A number of theories can be advanced from temperature changes, to improved habitat after a severe disease on the sea urchin (*Strongylocentrotus droebrachiensis*), to decreased predation of lobster early life stages by groundfish. The causes are likely to be multiple and interconnected. We simply do not have sufficient information on benthic communities to understand the changes that occurred.

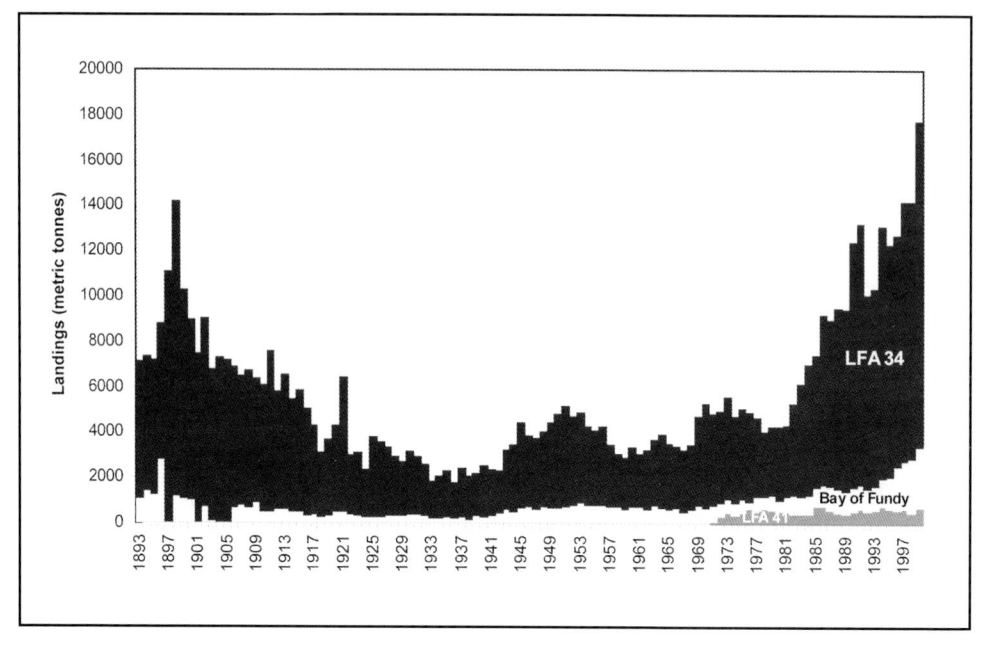

Figure 4: Lobster Landings 1893-2000

Interestingly, a recent comprehensive study (Fanning & Castonguay, 2003) aimed at explaining why collapsed fish stocks have failed to recover since the 1990s despite moratoria on commercial fishing, did come to the same conclusions. The publicly funded five-year study called 'The Comparative Dynamics of Exploited Ecosystems in the Northwest Atlantic Project' (CDEENA) did have enough data on oceanic conditions, fish populations and sea mammals abundance to study four Atlantic Canada ecosystems, namely Northern and Southern Gulf of St.Lawrence, Newfoundland and Eastern Scotian Shelf. It concluded that:

1) Ecosystems themselves appear to remain what they were, but that the key species did change substantially;
2) Most groundfish stocks have collapsed while both invertebrates and small pelagic fishes have increased;
3) Marine mammals now are the dominant cause of mortality on many

groundfish species (Morissette, Hammill & Savenkoff, 2003), and have replaced cod as the dominant predator in those systems.

Researchers involved also realised that all of their findings remain subject to high uncertainty, that an ecosystem approach demands massive amounts of information, and that further progress will have to address data gaps on benthos and foraging species.

Conclusions

At this time, when fish stocks are declining world-wide and climate changes are modifying ecosystems on which humans have depended for centuries as food sources, marine scientists have limited data sets generated by spurts of individual interest and funding.

Long-term monitoring programmes are poorly understood and generally undervalued by society. It follows that the interest of public institutions to systematically collect environmental data over long periods of time has generally been sporadic.

The scientific community has either not tried or has not been successful in selling the value of long-term observations to society. Strong rationales showing a crucial human interest transcending successive generations ought to be developed, sold to society and institutions by true leaders and visionaries, and henceforth supported by Institutions. There may, however, be a fundamental human problem here.

Decisions to initiate and to provide continuous support to long-term data collection are made by human beings in position of power. Unfortunately, there is a mismatch between the tenure of decision makers and the period of time over which data are required. Long-term data sets often transcend the life expectancy of scientists and politicians. A scientist's career spans 20–40 years; politicians see their mandates terminated or renewed every 3–7 years depending on countries.

Towards a new Vision

If this generation can muster the humility required for all learning, elements of an approach may come from First Nations and their Traditional Ecological Knowledge (TEK). TEK is based on the longest time series known to mankind. TEK is acquired through countless observations of nature by many successive generations of people who see themselves as part of the ecosystem that supports them.

TEK is formulated through a high level of abstraction capacity and adaptive intelligence. It is passed orally within family groups and from generation to generation by respected elders. The constant test of its veracity is the very survival of its holders. The inter-generational motive to maintain and constantly update TEK is survival and a deeply felt sense of duty to conserve resources for future generations.

At the beginning of this new Millennium, our collective challenge may well be to bring together ancient wisdom and the power of present day's science with its computers and satellites to find a way for our species to live within the carrying capacity of Planet Earth, our home.

Acknowledgements

The author is grateful to Cheryl Frail, Ken Frank, César Fuentes-Yaco, Elayne Myers, Jim Neilson, Linda Payzant and Trevor Platt for their valuable contributions, and specially to Ginette Robert for her suggestions and support. Elder Albert Marshall of the Eskasoni First Nation generously shared his knowledge and wisdom; his influence is humbly and gratefully acknowledged.

***René Lavoie** holds a PhD in estuarial ecology from Laval University in Canada. He has been involved in research and management of bivalves molluscan populations and molluscan aquaculture. He has consulted in fisheries research and aquaculture in the Caribbean, Asia and Africa. He maintains a keen interest in the traditional ecological knowledge of First Nations. He currently directs a research team on invertebrate fisheries at the Bedford Institute of Oceanography in Nova Scotia, Canada.*

3.15
SHELLFISH TOXICITY IN THE NW ATLANTIC:
UNEXPECTED AND WIDESPREAD OCCURRENCE OF *ALEXANDRIUM FUNDYENSE* BALECH IN COASTAL NEW ENGLAND, SEPTEMBER 1972. WHERE WAS THE MONITORING?

Christopher Martin
NOAA/National Marine
Fisheries Service,
Milford Laboratory,
Milford, CT 06460 USA

Introduction

THE sudden and exceptional proliferation and dominance of certain micro-algae in phytoplankton communities has always excited scientific interest. Visually prominent occurrences have sometimes been referred to as "red tides" owing to the discolouration of the sea water associated with large numbers of suspended pigmented cells or with the fluorescence emitted under conditions of intense solar illumination by cells near the sea surface. Special attention and study has been focused on events of this kind involving known toxic species. Harmful algal blooms, as the latter phenomena have come to be known, are now recognised as worldwide in occurrence. The presence of toxic organisms has frequently gone undetected until effects on organisms higher in the food chain including humans have been reported. Indigenous peoples along the eastern Pacific coast are said to have associated the appearance of nocturnal phosphorescence (more correctly, bioluminescence) with toxicity of shellfish in that region. Similarly, early explorers of the St. Lawrence region of Quebec, Canada reported that some native groups did not consume shellfish even though they were available in abundance.

Paralytic Shellfish Toxicity

The occurrence in late September, 1972 in coastal New England, of a spectacular and regionally widespread bloom of the toxic dinoflagellate *Alexandrium fundyense* Balech (originally thought to be *Gonyaulax tamarense* Lebour) marked the beginning of a renewed interest in the biological and chemical oceanography of the phenomenon. Moreover, owing to the pronounced neurotoxicity associated with this species, the public health implications were a driving force in stimulating widely diverse research efforts. The absence of a database on the occurrence and distribution of toxic dinoflagellates responsible for paralytic shellfish toxicity in the region hampered the understanding of what happened that year. It is significant in this conference setting to point out that long-term monitoring of these features of the environment was almost completely lacking. A brief description of the events of 1972 will help to place in proper perspective the response of natural resource and public health agencies to what was recognised as an emergency situation. Further, it may help to reveal how long-term monitoring with the compilation of appropriate data sets could have made a difference.

The realisation that coastal waters of New England from Maine to Massachusetts were suddenly and unexpectedly populated by extraordinarily large numbers of toxic microscopic aquatic plants emerged indirectly. Monitoring for such eventualities had not been seriously considered by fisheries or pubic health regulators and the scientific community was preoccupied with other things. The first public reports in local and regional news media were viewed with some suspicion on the waterfront: the alarming news was perceived as some new scheme by regulators and other officials to further hamper the pursuit of an honest living by clammers and other fisher folk. As it happened, the first observations that all was not well in the region did not come from laboratory scientists. Rather, managers of the Parker River National Wildlife Refuge, a preserve located on the northern coast of Massachusetts, first sounded the alarm. For several days the field staff there had been finding unusually large numbers of moribund and dead eider (*Somateria mollisima*) and black (*Anas rubripes*) ducks on the marshes and tidal streams within the refuge. Moreover, they noted the remains of fish and invertebrates stranded at the high tide line along the adjacent barrier beach. Exploratory necropsies at the refuge revealed nothing unusual. The birds, many of which used the refuge in their annual southward migration, had been feeding on blue mussels (*Mytilus edulis*), a favourite item of their diet. In attempting to explain the mortalities, officials at first speculated that a chemical spill had gone unreported and that the birds were succumbing to some, as yet, unidentified poison. The link to toxic algae had not been made.

At about the same time, bits and pieces of the puzzle were beginning to fall into place. Clammers were reporting odd behaviour of soft-shelled clams (*Mya arenaria*) freshly plucked from local mudflats: the siphons of harvested clams remained extended and flaccid as though paralysed. More worrying were the reports of restaurant patrons who checked themselves into local hospitals and doctors' offices with reports of a variety of symptoms from mild to severe paralysis of face and limbs to nausea and dizziness. They reported having consumed local shellfish. Medical staff found the symptoms puzzling as few were acquainted with the symptoms of shellfish poisoning. Patches of discoloured water were observed off nearby Cape Ann. An examination of samples ultimately revealed the dominance in high numbers of the toxic dinoflagellate, *A. fundyense*. Soon it was shown that aqueous extracts of local shellfish were found to be highly lethal to laboratory mice.

In response to these findings public health officials placed emergency bans on the harvesting and marketing of all molluscan shellfish from New England waters. While these steps were widely criticised as extreme, it was ultimately conceded that they were necessary to prevent serious illness. By the time this ban was imposed one victim of food poisoning had already required treatment in a hyperbaric chamber to relieve symptoms of respiratory paralysis. In addition to heightened public health concerns there were significant economic impacts on the shellfish industry at all levels. Shellfish toxicity in most commercially important bivalve molluscs remained above permissible levels for several months. At the time of this incident, resource agencies in Massachusetts and most neighbouring coastal states had no monitoring program in place aimed at detecting paralytic shellfish toxicity. Similarly, monitoring of toxic phytoplankton was absent from programmes throughout New England. Faced with this reality, new regulations including systematic monitoring of shellfish resource areas for the presence of paralytic toxins were put into place. Phytoplankton monitoring was regarded as too costly to implement so became the part-time interest of a few academic scientists.

While the reaction of public officials to this unexpected outbreak was effective in reducing public health impacts of toxin contamination of seafood and, to a lesser extent, the economic effects of a widespread ban on harvesting and marketing of affected seafood, it is clear that regulators were seriously handicapped. The recorded occurrence and distribution of the causative organism and the extent of its effects on resources was not available for the region in question. While significant help and advice was available from other jurisdictions within the United States and elsewhere (especially Canada), Massachusetts found itself faced with a problem that it understood poorly, if at all.

Going beyond immediate questions of public health and economics it seems appropriate to ask what such natural events might be telling us about the long-term health of coastal ecosystems, especially those dependent upon stable communities of phytoplankton and, in turn, the food webs dependent upon them.

Conclusion

What happened along the New England coast that autumn has not been satisfactorily explained. In the three subsequent decades since the outbreak, numerous attempts have been made to develop "predictive indices" in hopes of forecasting future plankton blooms. Most of these attempts have fallen far short of the mark. It is unlikely that successful results along these lines will ever be achieved in the absence of data sets derived from sustained detailed measurements of many of the features of such ecosystems. Ecosystem analysis based on extended time-series promises the best approach to understanding natural phenomena of this kind. Moreover, it may guide the direction of coastal ocean development involving such things as wastewater treatment, power generation, resource recovery, aquaculture siting and coastal land use. Work in the northwest Atlantic on harmful blooms has been fragmented and more a reflection of the vagaries of science funding or emergency responses to local events than it has to thoughtfully conceived and thoroughly conducted research. No coordinated effort has yet been made to establish long-term monitoring programmes in this region.

***Christopher Martin (retired 2005: Ed.)** received his PhD in biology at Harvard in 1962 and subsequently spent some time studying marine fungi in subtropical, tropical and Antarctic waters. His careers has also included stints in college teaching and for nearly 20 years has been in fisheries research with the US National Marine Fisheries Service. His interests have been focused on marine biotoxins and their impacts on shellfish aquaculture.*

3.16
CETACEANS: CAN WE MANAGE TO CONSERVE THEM?
THE ROLE OF LONG-TERM MONITORING

Greg Donovan
Head of Science,
International Whaling
Commission, The Red House,
135 Station Road, Impington,
Cambridge, CB4 9NP,
England, UK

INTRODUCTION

The order Cetacea comprises the whales, dolphins and porpoises. There are over 75 species, traditionally divided into the 'great' whales (see Table 1) and 'small cetaceans'[1]. Confusingly perhaps, small cetaceans include species called whales (e.g. killer whales, *Orcinus orca*) as well as dolphins (e.g. the white-beaked dolphin, *Lagenorhyncus albirostris*) and porpoises (e.g. the harbour porpoise, *Phocoena phocoena*). Cetaceans encompass a wide range of social structures and habitats.

Cetaceans are completely aquatic and are not easy to study for a number of reasons. As mammals, they need to return to the surface to breathe ('blow'), where they are briefly visible – the rest of the time they are out of sight. Dive times vary considerably by species and behaviour and 'long dives' can vary from up to 1–1.5hours (e.g. sperm whales, *Physeter macrocephalus*; northern bottlenose whales, *Hyperoodon ampullatus*) to 2–3 minutes (e.g. short-beaked common dolphins, *Delphinus delphis*; North Atlantic

---

[1] A full list of common and scientific names can be found at http://www.iwcoffice.org

right whales, *Eubalaena glacialis*). Typically animals blow several times after a long dive. However, even seeing cetaceans when they are present can be difficult, particularly in unfavourable weather conditions.

An additional difficulty, particularly for the migratory great whales, is that they can have extremely large ranges. For example, the eastern North Pacific population of gray whales (*Eschrichtius robustus*) migrates between summer feeding grounds in the Arctic to winter breeding grounds off Baja California, Mexico – some 7,500–10,000 km. Other baleen whale species and the sperm whale exhibit similar long migrations.

Table 1: The 'Great' Whales

Common Name	Scientific Name
Bowhead (or Greenland right whale)	*Balaena mysticetus*
North Atlantic right whale	*Eubalaena glacialis*
North Pacific right whale	*Eubalaena japonica*
Southern right whale	*Eubalaena australis*
Gray whale	*Eschrichtius robustus*
Blue whale	*Balaenoptera musculus*
Fin whale	*Balaenoptera physalus*
Sei whale	*Balaenoptera borealis*
Bryde's whale	*Balaenoptera edeni*
Common minke whale	*Balaenoptera acutorostrata*
Antarctic minke whale	*Balaenoptera bonaerensis*
Humpback whale	*Megaptera novaeangliae*
Sperm whale	*Physeter macrocephalus*

Management and monitoring

Whether we like it or not, humans have directly and indirectly influenced the environment of almost all species to a greater or lesser extent. Management can be said to be our attempt to limit and control the effects of humans on our environment, whilst obtaining the maximum 'benefit[2]' from that environment. In fact, everything we do (including, and perhaps especially, doing nothing) can be said to be a management decision. Given that, I believe that we have an obligation to try to ensure that we follow a wise and long-term management strategy. Although it might seem a semantic point, it should be remembered that we cannot 'manage cetaceans' – we can only manage human activities that may have an impact on cetaceans.

Human impacts on cetacean populations can be broadly classified into two groups: those that result in instantaneous or near-instantaneous death (e.g. direct hunting, incidental catches in fishing gear, ship strikes) and those that whilst not resulting in rapid death, affect the overall 'fitness' of the population (e.g. habitat related issues including pollution, overfishing of prey species, habitat loss). Whilst the impact of the first group will be clearly significant at the level of an individual, it may not be significant at the population level, depending on the number involved relative to the

---

[2]'Benefit' here can mean many things, ranging from direct exploitation to preserving pristine environment and abundance for future generation

total abundance. However, factors associated with the second group may not appear significant at the level of an individual but may be significant at the management stock level (e.g. individual females may appear healthy but if pollutant-induced physiological changes hinder their ability to reproduce, this may have long-term consequences for a population).

The first and most essential step in any management process is to *define the objectives* with respect to the status of the cetacean population(s) concerned. The second is to *assess the status* of those populations in the light of those objectives. The third is *determine management measures* that we think will ensure that those objectives are met and will continue to be met (in other words to identify and where necessary mitigate possible threats). The final and equally important step is *to monitor* the populations to make sure that the management measures are indeed working. It is important to note that the monitoring stage is not an optional extra – in an uncertain world it is essential that however perfect we think our management measures might be, we check to ensure that they are indeed working as we expect them to. Thus monitoring must be seen as an integral part of management, not an optional extra.

The purpose of this paper is to give a perspective on the need for long-term monitoring of cetaceans in a management context and how this might be achieved. Although scientific in nature, this is not a formal scientific paper. I have therefore not included references in the text but rather included a bibliography of useful reading at the end.

WHAT IS A MANAGEMENT PROCEDURE?

The concept of the management procedure approach was originally developed by the International Whaling Commission (IWC) Scientific Committee in the context of the management of direct catches (both aboriginal and commercial). However, the general approach is equally valid for other situations such as the management of incidental catches in fishing gear, the management of a reserve or protected area etc. In summary, the management procedure approach for direct or indirect catches is as follows (and see Figure 1):

1) agree management and conservation objectives, state them explicitly and assign them priorities;
2) agree and specify realistic data and analysis requirements;
3) accept scientific and practical limitations and take the inevitable uncertainty explicitly into account by determining a precautionary method of calculating catch limits involving rigorous testing via computer simulations for both quantitatively and qualitatively known sources of uncertainty;
4) after steps (1) – (3), adopt a management procedure that incorporates the process right through from data requirements and analysis to determination of catch limits (or other management advice);
5) include feedback monitoring to ensure that the agreed objectives are being met.

```
┌──────────────┐  ┌──────────────┐  ┌──────────────┐  ┌──────────────┐  ┌──────────────┐
│  OBJECTIVES  │  │    DATA:     │  │  PROCEDURES  │  │   CHOICE     │  │  IMPLEMENT   │
│              │  │ available and│  │  Scientists  │  │              │  │              │
│Users/scientists│ │  obtainable  │  │              │  │  Scientists  │  │  Scientists  │
│              │  │              │  │DESIGN FEATURES│ │    Users     │  │    Users     │
│   Managers   │  │  Scientists/ │  │Users/managers│  │              │  │              │
│              │  │    users     │  │              │  │   Managers   │  │   Managers   │
└──────────────┘  └──────────────┘  └──────────────┘  └──────────────┘  └──────────────┘
```

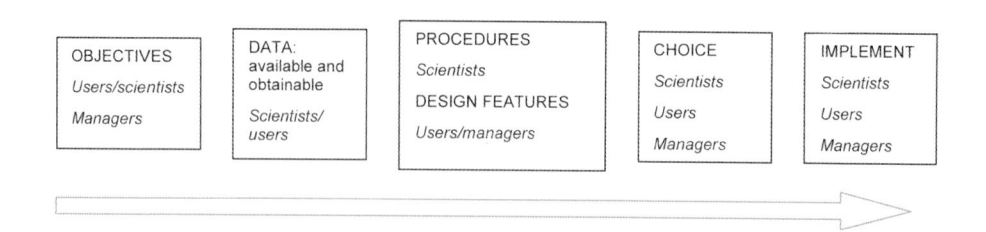

Figure 1: The steps to a management procedure.

An important feature of the process is that it should include all interested parties (stakeholders): scientists, managers and 'users'[3].

The advantages of such an approach are clear; everybody understands and agrees what are:

    1) the conservation and use objectives;

    2) the data requirements;

    3) the data analysis methods.

This removes the problems associated in the past with *ad hoc* assessment methods that could sometimes lead to greatly fluctuating scientific advice on appropriate catch levels from year to year. Such procedures are designed for long-term (decades) management. This allows *inter alia* appropriate long-term research planning. The users, managers, scientists and indeed the exploited populations, all therefore benefit from the management procedure approach.

In fact the process for developing a good monitoring programme alone (or within an overall management procedure) is very similar to that for developing the procedure itself (Figure 2). I will concentrate on this for the remainder of the paper.

```
┌──────────────┐  ┌──────────────┐  ┌──────────────┐  ┌──────────────┐  ┌──────────────┐
│  OBJECTIVES  │  │    DATA:     │  │  DETECTION   │  │    DESIGN    │  │  IMPLEMENT   │
│              │  │ available and│  │  CONFIDENCE  │  │              │  │              │
│Users/scientists│ │  obtainable  │  │              │  │  Scientists  │  │  Scientists  │
│              │  │              │  │  Scientists  │  │    Users     │  │    Users     │
│   Managers   │  │  Scientists/ │  │              │  │              │  │              │
│              │  │    users     │  │              │  │   Managers   │  │   Managers   │
└──────────────┘  └──────────────┘  └──────────────┘  └──────────────┘  └──────────────┘
```

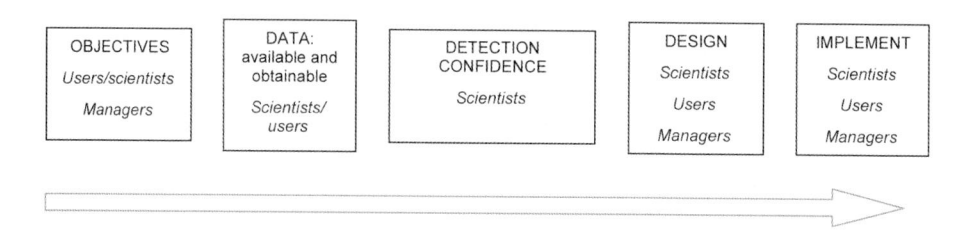

Figure 2: The steps to a monitoring programme.

OBJECTIVES

In attempting to develop a resource management or monitoring scheme, the most important initial step is to define management objectives. In effect, this means deciding

---

[3] By 'users' here, I include all interested parties – this may include whalers, fishermen, local communities, non-governmental organisations etc.

what we want to achieve by management and how we judge if it 'works'. It is relatively easy to arrive at 'extreme' objectives for the management of any natural resource:
- that the resource is not driven to extinction;
- that the maximum sustainable harvest is achieved.

It quickly becomes apparent however that within these two headings (state of the resource/needs of the user):
1) there are a wide number of options (e.g. see Table 2);
2) there has to be some trade-off between objectives.

For example the lowest risk of extinction occurs when there is no harvesting – this option, however, results in no chance at all of achieving the objective of the maximum sustainable harvest.

For cases where there is no desire to harvest a population, the objectives may be different but the principles remain the same. For example two objectives might be:
- maintain the population[4] at its present level;
- allow local communities to use the area in the traditional manner.

If it is found that the traditional activities (e.g. fishing) are a factor in the decline of the population, then again there will be a conflict between the objectives.

The setting of objectives and the relative weight given to those objectives (the trade-offs) ultimately require political rather than scientific decisions, although the scientist clearly has an obligation to explain the implications of any decisions that might be

Table 2: A few examples of possible management objectives

State of 'resource'
Prevent extinction
Keep greater than some percentage of its estimated pre-exploitation size
Keep at its 'current' level
Keep at some pre-specified target level
Return the population to its estimated pre-exploitation size
Maintain a particular trend in abundance
Keep at the level giving maximum productivity
Restore the distribution of the population
Maintain the habitat in its present state
Restore habitat to its original state
'Needs' of 'user'
Catch sufficient for operation
Maximum catch possible as soon as possible
Maximum catch possible eventually but allowing smaller catches to occur now
As quick a return on investment as possible
No effect on fishery (bycatches only)
Stable catches
Maintain sufficient numbers and distribution for whalewatching or tourism industry
To allow local communities to use the area in the traditional manner

[4] The question of what comprises a 'population', whilst extremely important, is beyond the scope of this paper.

taken to the politicians, for example by providing them with a range of specific options. If management is to be successful then it is extremely important that all interested parties (the 'stakeholders' as they are becoming known in the literature) are involved in the discussions leading to the setting of those objectives. There is a clear inter-relationship between objectives and monitoring – it is essential that any objectives related to the state of the resource are discussed in the light of our ability to be able to monitor whether we are meeting them.

CHARACTERISING STATUS

In this section I will briefly summarise some of the methods available to characterise the status of a cetacean population. The most common of these involves estimating the absolute abundance (and/or relative abundance) of the population and how that changes with time (i.e. trends in abundance/relative abundance). Which of these methods is appropriate will depend on the objectives set earlier.

Sightings methods

Dedicated shipboard or aerial surveys

This is not the place to explain in any detail how cetacean population size is estimated – that would take a book in itself and there are many excellent articles that deal with this subject. Needless to say, estimating the abundance of cetaceans is neither easy nor cheap, particularly if the aim is to estimate absolute rather than relative abundance. Up until now, the primary method for estimating cetacean abundance has been to use visual surveys and an approach called 'distance sampling'. In short, the idea is that although it is not possible to cover every square metre of a large area of ocean, with an appropriate survey design it is possible to estimate the density of animals along a 'strip' of ocean either side of a series of tracklines within the major survey area, and from this extrapolate the density to the whole survey area. Figure 3 shows one example of such a survey design.

The density of animals can be estimated by:

$$\frac{\text{The number of schools sighted } \tilde{} \text{ the mean school size}}{\text{The strip width } \tilde{} \text{ the total length of transect searched}}$$

This approach can be undertaken from vessels or from aeroplanes and major surveys, such as the North Atlantic Sightings Survey (NASS) that covered the Central and Eastern Atlantic, or the Small Cetacean Abundance in the North Sea (SCANS) survey that covered the North Sea and adjacent waters, often use both. As with any such approach, there are a number of assumptions about the method that may be violated and thus any estimate of abundance (and series of abundance estimates in the case of monitoring) will have considerable uncertainty surrounding it. For example, this can arise from a number of causes ranging from differences in the abilities of observers,

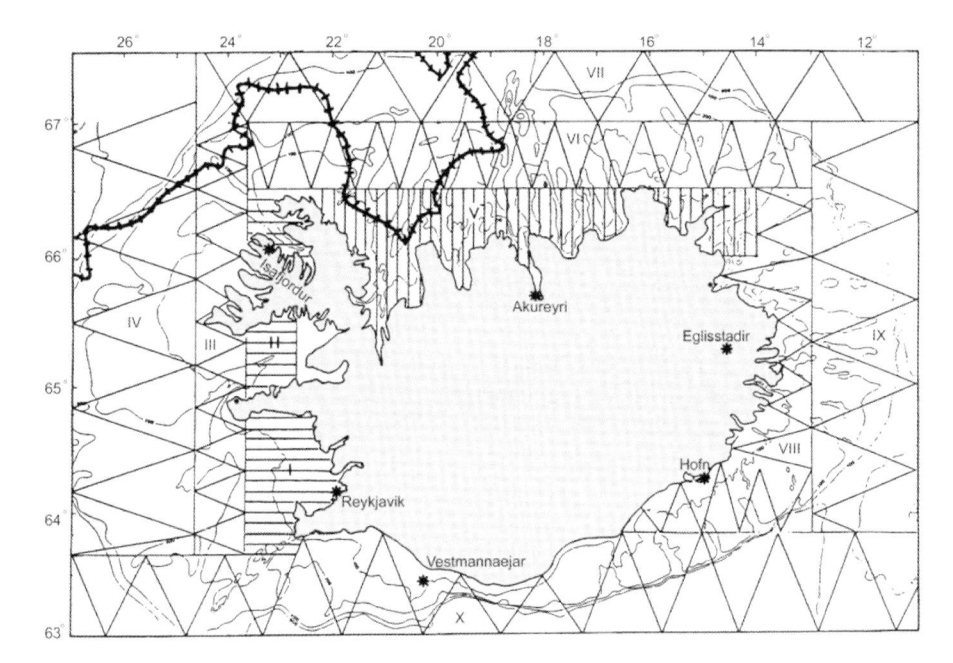

Figure 3:. Examples of track design from an aerial survey around Iceland

differences in weather conditions and problems in data collection, to differences in the distribution of animals over time for example due to changes in prey distribution due to oceanography, ice cover, etc (Table 3). It is important to try to collect data to allow a quantitative investigation of these factors. As discussed below, these are also all

Table 3: A few examples of some of the factors affecting the sighting of cetaceans. Whatever visual survey approach is adopted, it is important to collect data that enable these factors to be quantified.

Some factors affecting sightability even if animals are present
The ability (and mental state!) of the observer
The number of observers
The amount of time they spend searching
The nature (e.g. height, stability) of the sighting platform
The weather/sea conditions
The size of the animals
The group size of the animals
The behaviour of the animals

Some factors affecting whether animals are present in the survey area
Time of the year (if undertake seasonal migrations/movements)
Oceanographic factors (e.g. water temperature, salinity, currents)
Ice cover
Distribution of prey species (in suitable concentrations)
Distribution of competitors (including humans, e.g. vessels)

important in the context of the ability to determine trends. Good guidelines for such surveys can be found at www.iwcoffice.org.

In essence, although the theoretical and practical aspects of visual surveys have improved greatly in recent years, the limitations of the results must be recognised. What is obtained from a single survey is a 'snapshot' – an estimate of the number of animals in a particular geographical area at a particular time. Interpretation of such results requires knowledge of the relationship of that area and time to the 'behaviour' of the population. The collection of additional data either during surveys or from other simultaneous sources (e.g. with respect to oceanography, bottom topography, productivity etc.) to try to explain the distribution of cetaceans within the survey area is becoming increasingly common and important. A powerful argument for monitoring programmes is that a single 'snapshot' however well obtained can be misleading. There are plenty of examples of cetacean populations changing their distribution dramatically from one year to the next.

The question of using data collected on dedicated surveys to monitor populations is discussed under the 'Trends' section below.

Opportunistic vessel surveys

The cost of the vessel is clearly a major factor in the cost of surveying. It is tempting therefore to look at the possibility of 'piggy-backing' onto other vessels – either by using sightings information they submit or by placing observers on them. I will deal with each of these in turn.

Incidental sightings from platforms of opportunity

The most obvious advantage of such an approach is that data collection is essentially free; a further advantage can be that it does involve a variety of 'stakeholders'. However, there are a number of quite serious disadvantages that preclude the use of such data for reliably estimating abundance and relative abundance. These can be summarised as follows:

- lack of control as to the area and tracklines covered;
- lack of control over the timing of the effort
- lack of ability to estimate 'effort'[5];
- lack of control over the experience of the observer;
- lack of quality control – e.g. with respect to species identity, group size and other basic data.

All of this means that the data are of limited value. Despite this, a number of programmes (such as some based in the UK[6]) exist that do their best to take into account the above problems by providing advice and training materials and using stringent conditions before accepting the data. At the very least, a well-designed programme can

---

[5]This is essential if information is to be used in any quantitative way. As a minimum one would need to know how much time was spent on concentrated searching and what the weather and sea conditions were.
[6]E.g. see http://www.jncc.gov.uk/Publications/cetaceanatlas/default.htm

give some information on occurrence and distribution throughout a broad area and time and provide assistance in designing dedicated surveys.

Dedicated sightings from platforms of opportunity

The next stage 'up' is where dedicated observers are placed on vessels which routinely operate in specific waters, often following specific routes. Examples of such vessels include research vessels, whalewatching vessels, coast guard vessels and ferries. The amount of information that can be obtained from such an approach is dependent *inter alia* on the nature of the vessel (size, speed, viewing platform, noise etc), the cruise tracks and the ability to divert from the track to confirm species and group size. In certain 'ideal' circumstances, data can be used to estimate density and even absolute abundance (e.g. this is the case for a collaborative programme in the Antarctic where cetacean observers participate on research vessels examining krill abundance[7]), in other cases sightings from ferries have been used to investigate distribution and relative abundance at different times of the year. However, where there is no control over the tracklines careful consideration must be given as to whether the results are representative of the population.

Once again, the individual circumstances will need to be examined in order to assess the suitability of the resultant data with respect to information on abundance and distribution.

The use of such data to monitor populations is discussed under the 'Trends' section below.

Land-based censuses[8]

Under certain relatively rare circumstances, it is possible to obtain absolute abundance estimates from shore. A major advantage of a shore-based census is that it is relatively cheap. It is also easy to collect all the necessary associated data (see Table 3). The two best examples are for the Bering-Chukchi-Beaufort Seas bowhead whales and the eastern North Pacific gray whales. In both cases the animals have a very narrow migratory corridor such that at a certain point along the migration route, all the animals pass within sight of the shore under ideal conditions. Although corrections need to be made for various factors, including weather conditions, animals passing beyond the range of the observers, night time migrations, the estimates produced from such censuses have produced some of the best time series of abundance estimates (see below).

Land-based/fixed base systematic observations

A major advantage of land-based surveys (or fixed-base platforms such as oil rigs) is that they can be cheap. It is also easy to quantify effort and other factors (see Table 3). Their most obvious limitation is the very small area that can be covered by the observers. Even with several 'stations' along a coast, observations will be limited to

---

[7] See http://www.ccpo.odu.edu/Research/globec/iwc_collab/menu.html
[8] The word 'census' is used in this section because it is theoretically possible to count all of the animals in the population, unlike traditional distance-based surveys where abundance is estimated by extrapolating from a sampled area to the whole area.

coastal waters. Considerable care is thus needed in interpreting the data. It is clearly not appropriate to use this method to reach conclusions on the abundance of species with oceanic distributions. Care is also needed in making inferences about coastal populations – even a small change in distribution can lead to a false interpretation. However, they can produce interesting information on the distribution and relative occurrence of coastal species through time.

Mark-recapture methods

This is a relatively common way to estimate abundance of animal populations. The simple idea is that you 'capture' and mark a sample of animals and then release them. You then go out and capture another sample of animals and see how many of them are those you marked the first time. The ratio of marked to unmarked animals in the second sample is assumed to be equal to the ratio in the population at large. It is thus easy to calculate the total population size as:

$$\frac{\text{The number of animals in the first sample} \times \text{the number of animals in the second sample}}{\text{The number of marked animals recaptured}}$$

Of course, nothing in life is that simple and there are a number of assumptions that are made that can never be completely fulfilled. It is beyond the scope of this paper to discuss them in detail but they include: the marked animal will always be recognised if it is caught; the marked animals are representative of the population; marking an animal does not make it less likely to be caught again; and each animal in the population has the same chance of being captured during any sampling occasion. As with the visual survey approaches, it is important that data are collected to allow one to see if the assumptions are being violated and if so, by how much.

There are two types of 'marks' that are used in cetacean studies. The most common is the use of photographic 'capture' of individual marks such as the colour pattern and outline of the flukes of humpback whales (see Figure 4) or the nicks and scars on the dorsal fins of bottlenose dolphins (see Figure 5)[9] – these are all individually different, rather like fingerprints in humans. The second, more recent approach has been to obtain small biopsy samples from free-swimming animals and then to analyse them to obtain their genetic 'fingerprint' – again a technique now commonly used in criminal investigations to identify individuals.

Such data can and have been used to estimate both abundance and trends in abundance of animals. For example a major study of humpback whales in the North Atlantic used both photographic and biopsy 'marks' to obtain an abundance estimate for the entire western North Atlantic of about 11,600 (95% confidence interval 10,100 to 13,200) in 1992/93. An annual increase rate for the Gulf of Maine humpback whales for the period 1979–1993 was estimated at about 3%. Another example of this approach being used is to estimate the abundance of bottlenose dolphin populations in various parts of the worlds including Ireland[10].

---

[9]Not all species have easily recognisable natural marks.
[10]http://www.shannondolphins.ie/

Figure 4: Examples of humpback whale fluke photographs suitable for individual identification studies. Note that the two bottom photographs are of the same whale on different days.
Photographs courtesy of the author).

Figure 5: Examples of bottlenose dolphin fins suitable for individual identification studies, showing scars, nicks and colour patterns. *(Photographs courtesy of Ana Cañadas, Alnitak, Spain).*

Acoustic methods

Many cetaceans produce sounds. The potential to be able use underwater sound detections to estimate absolute or relative abundance is attractive and considerable progress has been made in developing techniques to achieve this either using towed hydrophones (which can be used from both dedicated and 'opportunistic' vessels) or fixed hydrophones that can be attached to the sea-floor, particularly for harbour porpoises and sperm whales. Some advantages of such approaches are that they can be automated and can be used under poor weather conditions, but further work is required to try to overcome disadvantages such as the requirement for animals to be vocal and to try to determine the link between vocalisations and abundance. A combination of acoustic and visual surveys seems particularly promising. One specialist and elegant application of this is the use of acoustic location data to detect bowhead whales that migrate under the ice when passing the visual observers in the census referred to earlier.

TRENDS

Can we believe them?

Estimating trends in abundance at its simplest involves comparing two or more estimates of absolute or relative abundance. The most important aspect of any such comparisons is to determine whether or not the estimates are truly comparable. The key word is here is consistency. It is important that great care is taken to ensure that to the extent possible, surveys are carried out using the same methods and equipment, the same area and the same time of year. If 'improvements' are to be made in a series of surveys, it is extremely important that such changes are examined, not simply to see if they result in a 'better' estimate of absolute abundance but also whether the changes might compromise the interpretation of the series of results.

Where the absolute or relative abundance estimates refer to small areas, they may be of limited value in estimating trends. Relatively small changes in distribution (not unlikely in highly mobile animals such as cetaceans) can have a large impact on abundance estimates, making interpretation of apparent trends extremely difficult. Choice of an appropriate survey area is thus of great importance.

Even with well-designed surveys, interpretation of long time series can be difficult. A good example of this is the series of IWC Antarctic cruises to estimate Antarctic minke whale abundance undertaken since 1978. It is not possible to survey the Antarctic in one year and so parts are covered each year. Between 1978/79 and 2003/2004, the complete 'circumpolar' survey was achieved three times. Over that period, inevitable changes occurred with respect to crew, vessels, survey design, and environmental conditions (such as ice cover). Despite considerable effort it is proving extremely difficult to determine whether observed changes in *estimated* abundance reflect real changes in abundance, changes due to methods and personnel, changes in distribution or a combination of these.

An understanding of the reason why animals are distributed the way they are (e.g. by relating distribution and abundance to environmental factors) can prove a valuable tool in interpreting trend data; recently developed methods of 'spatial modelling' that attempt to do this are very promising.

A simple hypothetical example is given in Figure 6. If there had been a change in methods after year 20, then it would be difficult to state whether the observed increase in the estimate was real or merely an artefact of the changed methodology. This should not be taken as implying that methods should never change or be improved – rather it is a warning to ensure that if changes are thought to be necessary, it is important that the implications for a long-term data series are taken into account. For example, where possible the two methods should be used in parallel for a time to ensure compatibility.

Is it realistic to manage based on trends?

It should be clear by now that estimating abundance of cetaceans is difficult. Even with a well-designed survey in good weather, the uncertainty in an abundance estimate will be quite large. This makes determining a statistically significant change in population size difficult. If a proposed management objective is to manage a population based on a

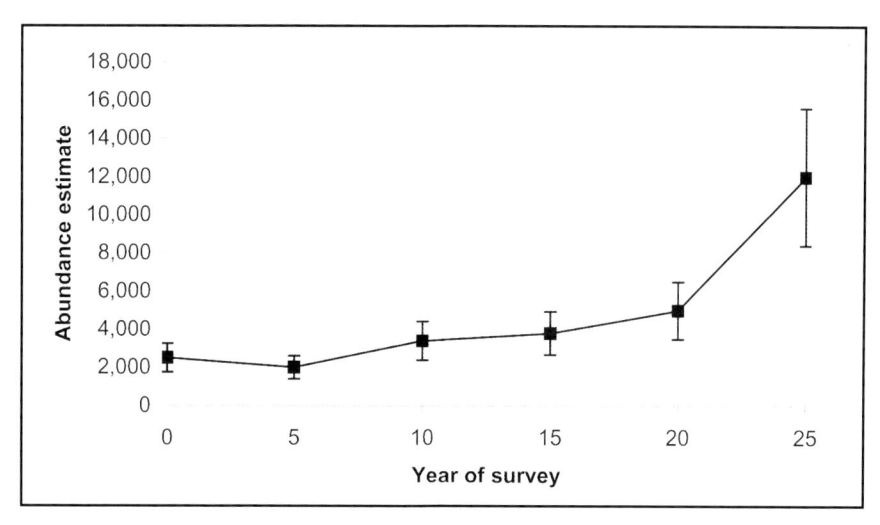

Figure 6: Hypothetical example of the results of abundance surveys over a 25 year period (see text).

characterisation of status that involves trends (e.g. to keep a population stable) then it is essential that you test whether it is possible to estimate a trend (or be sure that there isn't one) using your expected methods in a realistic time frame. It can take decades to detect changes in abundance. Without going into details, it is important to carry out a 'power analysis' to test this. In effect the approach is to assume that you will obtain an estimate with a particular level of confidence and then see how many abundance estimates carried out at a particular frequency will allow you determine whether the population has changed at a chosen rate (e.g. 10%, 20%, 50%). After examining the results of such an analysis, it may be necessary to adopt different objectives.

A conservative and precautionary approach to management has been adopted by the International Whaling Commission. In essence, it uses computer simulations of whale populations to choose an approach to setting catch limits that will still be safe even if the inevitable scientific uncertainty surrounding abundance estimates and trends (as well as many other factors) is taken into account.

DESIGNING A MONITORING PROGRAMME

It should be noted that the development of cost-effective monitoring programmes to determine whether mitigation measures for threats to cetaceans (e.g. bycatches) are successful is the subject of a number of ongoing research programmes – it is not a simple matter. The design of a monitoring programme must take into account all of the features discussed above – there is no single answer – it will depend on a number of factors and will almost inevitably be case-specific. The important thing is to develop a monitoring programme that will meet the needs of the management objectives chosen in a cost-effective manner and one which it is feasible to carry out at regular intervals over a long period. In many cases, a combination of methods will be appropriate. Each of the options should be examined for its strengths and weaknesses in the context of the chosen objectives. For example, large-scale surveys are extremely expensive. It may be that undertaking such surveys at regular but longer intervals (e.g. 10 years) in combination with more regular, cheaper smaller scale local surveys is appropriate. Acoustic approaches hold considerable promise but require further development. The use of computer simulations to test 'what if we get it wrong?' is an effective and relatively cheap way to determine if a programme will be effective – it can prevent unnecessary resources being expended on uninterpretable research and most importantly, the potential consequences for the animals are far less severe.

CONCLUSIONS

1) Long-term monitoring programmes are an essential part of management – not an optional extra.
2) Involve stakeholders at all stages of developing a management strategy.
3) It is essential to define management objectives – without these it is not possible to develop an appropriate monitoring programme.
4) Weigh up the needs of monitoring whether the management objectives are being met with the advantages and disadvantages of the various ways of estimating the abundance and distribution of the target cetacean populations – the use of power analyses and computer simulations cannot be over-emphasised.
5) Where practical, collect data that will enable a better understanding of why animals are where they are when they are – it is essential for properly interpreting trend data.

BIBLIOGRAPHY AND ADDITIONAL READING

The interested reader is referred to the following publications which are given in full in the reference list at the end of this book. :

Buckland & Breiwick (2002)
Buckland *et al.* (2001)
Evans & Hammond (2004)
Garner *et al.* (1999)
Gerrodette (1987)

Hall & Donovan (2002).
Hammond (1986)
Hammond & Donovan (In press)
Hammond *et al.* (1990)
Hammond *et al.* (2002)

Hiby & Hammond (1989)
Matsuoka *et a.l* (2003)
Palsbøll *et al.* (1997)
Raftery & Zeh (1998)
Smith *et al.* (1999)

Greg Donovan has worked with the whales, dolphins and porpoises for 25 years and is now head of science at the International Whaling Commission and also edits the Journal of Cetacean Research and Management. His major interests involve population biology and modelling to try and provide unbiased scientific advice on the conservation and management of these fascinating animals. His fieldwork has taken him to Greenland, Iceland, Norway, Spain, Alaska and Peru, in each place leaving behind groups of people with a love of Irish music and Guinness!

3.17
ENVIRONMENTAL MONITORING IN IRELAND: ASPECTS OF THE ROLE OF THE ENVIRONMENTAL PROTECTION AGENCY

Larry Stapleton
Environmental Protection Agency,
PO Box 3000, Johnstown Castle
Estate, Wexford,
Ireland

Introduction

An overview is presented of the role of the Environmental Protection Agency (EPA) in relation to environmental monitoring in Ireland. The central part of this role is the preparation of national environmental monitoring programmes. In this regard, an updated national monitoring programme for tidal waters has been prepared. The main features of monitoring programme are outlined, with particular reference to the Water Framework Directive (2000/60/EC). Some results from the monitoring of rivers and tidal waters, are presented, including some key results that illustrate the importance of long-term monitoring. Reference is made also to the results of a recent assessment of the predicted impacts of climate change in Ireland – a further factor that underlines the importance of long-term monitoring.

Legislative Requirement

Under Section 65 of the Environmental Protection Agency Act, 1992, the EPA is charged with developing and publishing national monitoring programmes for the environment, and with taking appropriate steps to ensure that the programmes are implemented. The Act defines "monitoring" to include "the inspection, measurement, sampling or analysis for the purposes of this Act of any emission, or of any environmental medium in any locality, whether periodically or continuously". The EPA is required to specify:

 a) the nature and extent of the monitoring to which the programme relates and the reasons why, in the opinion of the Agency, the monitoring should be carried out;

 b) the persons or bodies (including the Agency) by which the intended monitoring is to be carried out;

 c) the resources, including equipment, other facilities and staff, required to carry out the monitoring and the cost thereof;

 d) the arrangements which the Agency considers appropriate for access to, dissemination of, and publication of the results of the monitoring.

Monitoring Programmes

In accordance with the 1992 Act, national monitoring programmes for various aspects of the environment have been produced. These include the following:

- Estuarine and Coastal Water Quality (1996)
- Groundwater Quality (1997)
- Air Quality (2000)
- Lake Water Quality (2001)
- River Water Quality (2002)

In addition, a National Hydrometric Monitoring Programme has been prepared, covering the measurement of flows and water levels in rivers and lakes.

National Environmental Monitoring Programme for Transitional, Coastal and Marine Waters (2003)

The more recent monitoring programmes have been prepared in the context of the Water Framework Directive (WFD). It was recognised that the 1996 monitoring programme for estuarine and coastal waters needed to be updated to take account of this Directive and of other recent developments, and the new programme has now been completed. In line with the terminology used in the WFD, 'transitional' replaces 'estuarine' in the title of the new monitoring programme. The document was published initially as a discussion document.

The national monitoring programme for transitional, coastal and marine waters is a programme for the environmental monitoring of tidal waters to be undertaken by the relevant public authorities in the state over the coming years. The principal bodies involved in its preparation have been as follows:

 Environmental Protection Agency
 Marine Institute
 Radiological Protection Institute of Ireland
 Met Éireann
 National Parks and Wildlife

Input was received also from the Geological Survey of Ireland, the Central Fisheries Board, the Department of the Environment, Heritage & Local Government, the Department of Communications, Marine & Natural Resources and from local authority representatives.

This range of organisations involved in the process reflects the diversity of responsibilities for the monitoring of the environment of tidal waters, and indeed within each organisation there is a variety of types of monitoring undertaken. The development of this monitoring programme provides a unique overview of existing and proposed monitoring of tidal waters in Ireland.

The overall objective of the new national monitoring programme is to enable the achievement and maintenance of the integrity of biotic communities and habitats in transitional, coastal and marine waters, and to ensure protection of the beneficial uses of these waters and the protection of human health in relation to these uses. Other objectives include the following:

- providing information on the quality of the estuarine, coastal and marine environment that will allow for ongoing assessment of its overall condition and
- providing a means of identifying and tracking trends in water and environmental quality.

Specific obligations for monitoring and related matters arise in respect of the Water Framework Directive, the Joint Assessment and Monitoring Programme (JAMP) of the Oslo & Paris Convention (OSPAR), and principal national legislation. The preparation of the new programme has presented an opportunity to review the adequacy of existing programmes, and to introduce as appropriate additional measures arising from these and other new requirements.

The Water Framework Directive (WFD) in particular is bringing about a significant shift in the approach to the management of water quality. The purpose of this Directive is to establish a framework for the preservation and, where necessary, the improvement of water quality of inland surface waters, transitional and coastal waters and groundwater. The overriding objective of the WFD is to achieve at least good water status by 2015.

The Directive contains obligations to carry out a very large number of tasks in a variety of areas, including scientific/technical, information management, economic and administrative. Of particular relevance is the WFD requirement to implement monitoring programmes by December 2006.

Article 8 of the Directive requires programmes to establish "a coherent and comprehensive overview" of water status. Monitoring is required to establish the status of water bodies identified as being at risk of failing to achieve their environmental objectives, and to assess any changes in the status of such water bodies resulting from the programmes of measures to be introduced under the Directive.

Monitoring programmes under the WFD comprise three main elements:

- *A Surveillance Monitoring Programme.* This is an extensive programme intended to provide an overall assessment of water status. It is repeated at relatively lengthy intervals. The normal cycle corresponds to the six-year River Basin Management Plan (RBMP) cycle. The results are to be used to assess long-term natural variability and to assess changes resulting from widespread anthropogenic activity.

- *An Operational Monitoring Programme.* This is a more intensive programme intended for application in those water bodies that are, according to the initial characterisation at risk of failing to meet their environmental objectives. These will include water bodies that are (or are at risk of being) classified as of 'moderate', 'poor' or 'bad' status, or water bodies of 'high' status or 'good' status that are at risk of losing this status. It is also intended to be applied to bodies of water into which priority list substances are discharged.
- *An Investigative Monitoring Programme*, intended for use as necessary to identify the causes of accidental pollution or other causes of failure to achieve environmental objectives.

The National Environmental Monitoring Programme for Transitional, Coastal and Marine Waters describes the monitoring roles, existing and proposed, of national, regional and local bodies in relation to each of the thirty-six separate components of the Programme (Diagram 1). The document details the objectives, monitoring approach and reporting arrangements proposed for each of the component programmes, which are organised in six major themes:

      A **Physical Aspects (Physical Oceanography and Meteorology)**
      B **Ecological Integrity and Biodiversity**
      C **Water Quality and Trophic Status**
      D **Hazardous Substances**
      E **Food Safety and Human Health**
      F **Radioactive Substances**

A Marine Monitoring Forum will oversee the implementation of the Programme on an ongoing basis, to be composed of representatives of the relevant Government Departments, national and regional bodies and local authorities. This body will review the adequacy of existing monitoring arrangements, recommend research into new methods and approaches and make recommendations for modification as it considers necessary the programme.

Monitoring of Activities Licensed by the EPA

The EPA licensing function includes the licensing of activities that fall within the scope of EU Integrated Pollution Prevention and Control (IPPC) Directive (96/61/EEC) and of the Waste Management Act. The Office of Environmental Enforcement, within the EPA, is directly responsible for the enforcement of EPA licences issued to industrial, waste and other activities. In addition, a National Environment Enforcement Network has been established by the EPA with the objective of protecting and improving the environment by the effective implementation and enforcement of environmental laws.

Structure of the National Environmental Monitoring Programme for Transitional, Coastal and Marine Waters

A — Physical Aspects
- (A.1) Physical Oceanography Observation Programme
- (A.2) Marine Meteorological Observation Programme
- (A.3) National Seabed Survey
- (A.4) Hydromorphology in Transitional and Coastal Waters (WFD)

B — Ecological Integrity and Biodiversity
- (B.1) Conservation Status (Natura 2000)
- (B.2) Status of Commercial Fish Stocks
- (B.3) Impacts of Fishing on Non-Target Organisms
- (B.4) Environmental Monitoring of Waste Management Sites
- (B.5) Environmental Monitoring of Aquaculture Sites
- (B.6) Status of Phytoplankton in Transitional and Coastal Waters (WFD)
- (B.7) Status of Macrophytes in Transitional and Coastal Waters (WFD)
- (B.8) Status of Macrobenthos in Transitional and Coastal Waters (WFD)
- (B.9) Status of Fish in Transitional Waters (WFD)
- (B.10) Status of Introduced Non-Indigenous Species

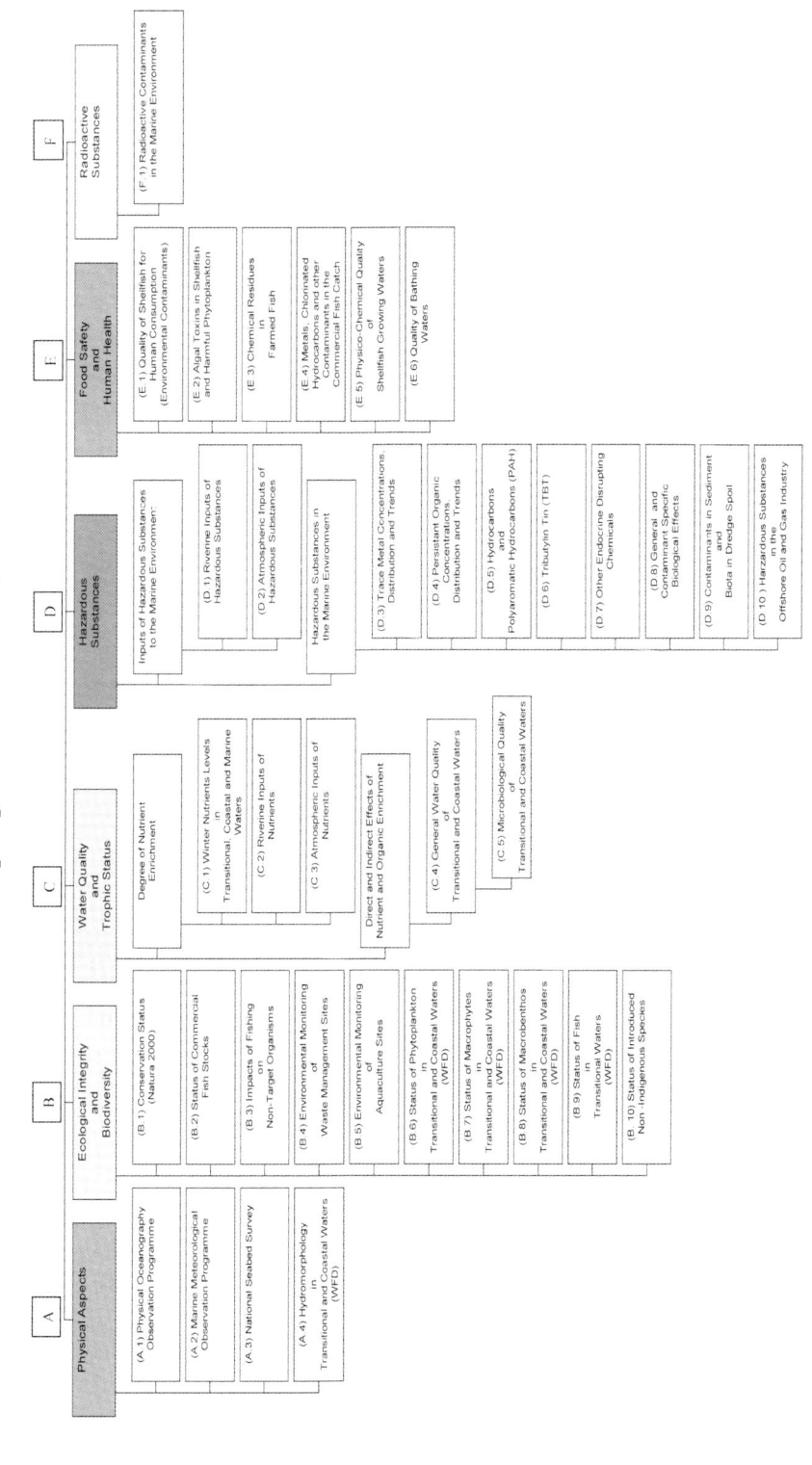

C — Water Quality and Trophic Status
- Degree of Nutrient Enrichment
  - (C.1) Winter Nutrients Levels in Transitional, Coastal and Marine Waters
  - (C.2) Riverine Inputs of Nutrients
  - (C.3) Atmospheric Inputs of Nutrients
- Direct and Indirect Effects of Nutrient and Organic Enrichment
  - (C.4) General Water Quality of Transitional and Coastal Waters
  - (C.5) Microbiological Quality of Transitional and Coastal Waters

D — Hazardous Substances
- Inputs of Hazardous Substances to the Marine Environment:
  - (D.1) Riverine Inputs of Hazardous Substances
  - (D.2) Atmospheric Inputs of Hazardous Substances
- Hazardous Substances in the Marine Environment
  - (D.3) Trace Metal Concentrations, Distribution and Trends
  - (D.4) Persistent Organic Concentrations, Distribution and Trends
  - (D.5) Hydrocarbons and Polyaromatic Hydrocarbons (PAH)
  - (D.6) Tributyln Tin (TBT)
  - (D.7) Other Endocrine Disrupting Chemicals
  - (D.8) General and Contaminant Specific Biological Effects
  - (D.9) Contaminants in Sediment and Biota in Dredge Spoil
  - (D.10) Hazardous Substances in the Offshore Oil and Gas Industry

E — Food Safety and Human Health
- (E.1) Quality of Shellfish for Human Consumption (Environmental Contaminants)
- (E.2) Algal Toxins in Shellfish and Harmful Phytoplankton
- (E.3) Chemical Residues in Farmed Fish
- (E.4) Metals, Chlorinated Hydrocarbons and other Contaminants in the Commercial Fish Catch
- (E.5) Physico-Chemical Quality of Shellfish Growing Waters
- (E.6) Quality of Bathing Waters

F — Radioactive Substances
- (F.1) Radioactive Contaminants in the Marine Environment

Diagram 1

The members of the network include representatives from local authorities, state agencies and Government Departments that are involved in enforcement of environmental legislation.

Ongoing long-term monitoring of the licensed activities is a key part of the enforcement function, and this includes measurement of site based performance by operators against licence conditions, based on emission limits designed to ensure environmental protection and performance standards. Monitoring may include sampling and analysis of emissions and of environmental media. In addition to the requirement for the licensees to undertake monitoring, the EPA carries out regular independent monitoring (or in some cases has it undertaken on its behalf). The EPA conducts audits of licensed facilities to determine the extent of the compliance with the licence, the progress of implementation of agreed programmes of work and to evaluate the licensee's environmental management system. Thus, the major industrial, waste management, and related activities in the country are subject to a rigorous programme of long-term monitoring.

Selected Monitoring Results

Tidal Waters

The monitoring programme for estuarine and coastal and waters published in 1996 had a strong focus on guiding the implementation of two key EU Directives in Ireland, namely those on nitrates from agricultural sources (91/676/EEC) and on urban waste water treatment (91/271/EEC). Detailed surveys of the main tidal waters receiving inputs from urban and riverine sources had been underway from 1995.

The data for the five-year period 1995–1999 were examined and reported by the EPA (2001). A five-year period was considered to be the minimum necessary to provide sufficient information on which to base an assessment of trophic status. In accordance with the definition of eutrophication in these Directives, a set of quantitative salinity-related and linked criteria was developed comprising:

      (a) criteria for enrichment;

      (b) criteria for 'accelerated growth;'

      (c) criteria for 'undesirable disturbance.'

The assessment of a number of areas was updated subsequently based on information available to 2002. Overall these assessments have led to the following tidal waters being classified as eutrophic:

          Broadmeadow Estuary Inner

          Liffey Estuary

          Slaney Estuary Upper & Lower

          Barrow Estuary

          Suir Estuary Upper

          Owennacurra Estuary/North Channel.

          Bandon Estuary Upper & Lower

Lee Estuary Upper (Tralee)

Feale Estuary Upper & Cashen/Feale

Killybegs Harbour.

In addition, the following waters were classified as potentially eutrophic:

Castletown Estuary

Blackwater Estuary Upper & Lower

Lee Estuary/Lough Mahon

The monitoring programme in respect of trophic status is ongoing and will track progress with the implementation of EU requirements in the coming years.

Rivers

The EPA's biological monitoring programme for rivers is a continuation of a countrywide programme that has been in place since the early 1970s. Records are available since 1971 from a baseline of stations representative of 2900 km of Irish river channel. The data for these baseline stations provide a valuable indication of long-term trends in river water quality in Ireland (Figure 1).

A larger baseline of 13,200 km, representative of all the major river channels, has been monitored over more recent decades and provides the trends in river water quality, representative of the main rivers for the country as a whole (Figure 2).

The most recent review period was 1998–2000 (McGarrigle et al., 2002) and the overall quality situation in the 13,200 km of freshwater channel surveyed was as set out in the Table 1.

Table 1: Trends in water quality of major river channels in Ireland, 1998–2000.

Water Quality Class		km (%)
Class A	(Unpolluted)	9237 (70)
Class B	(Slightly Polluted)	2257 (17)
Class C	(Moderately Polluted)	1637 (12)
Class D	(Seriously Polluted)	112 (1)

The data show that there was a slight improvement in the position compared to the 1995–1997 period. The proportion of channel in Class A increased from 67 to 70 per cent with consequent reductions in the channel lengths in Classes B and C. The length of seriously polluted channel remained at around 1 per cent. The increase in Class A was also observed in the 2900 km long-term baseline. Most of the improvements were seen in the Shannon and South Eastern Regions. Thus, for the first time since national reports were prepared in the early 1970s, a reversal of the trend to greater deterioration has been recorded.

While there may be several other factors involved in bringing about this change, it is probably due, in part at least, to the impact of the Government's counter-eutrophication strategy. In particular the building of new sewage treatment plants incorporating phosphorus removal, and the various measures introduced to address nutrient losses

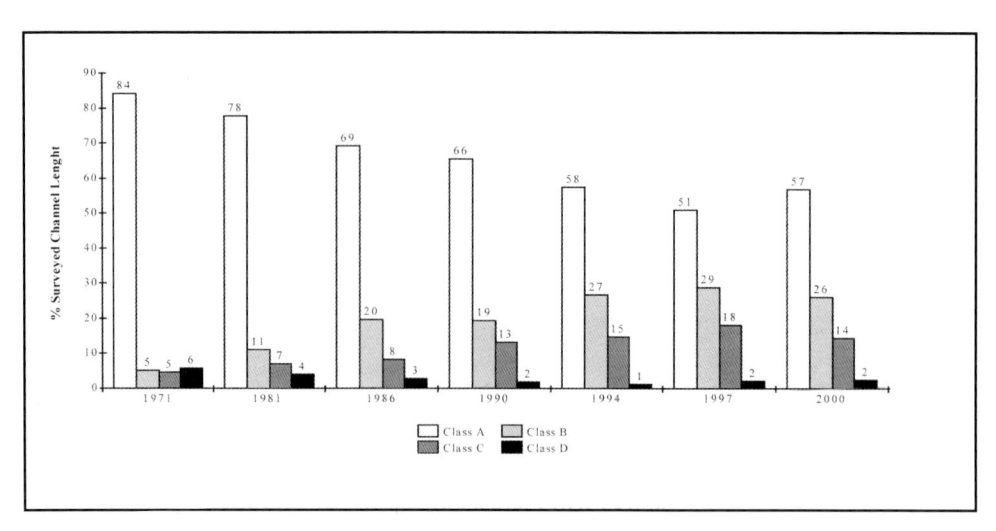

Figure 1: Long-term trends in river water quality (2900 km baseline)

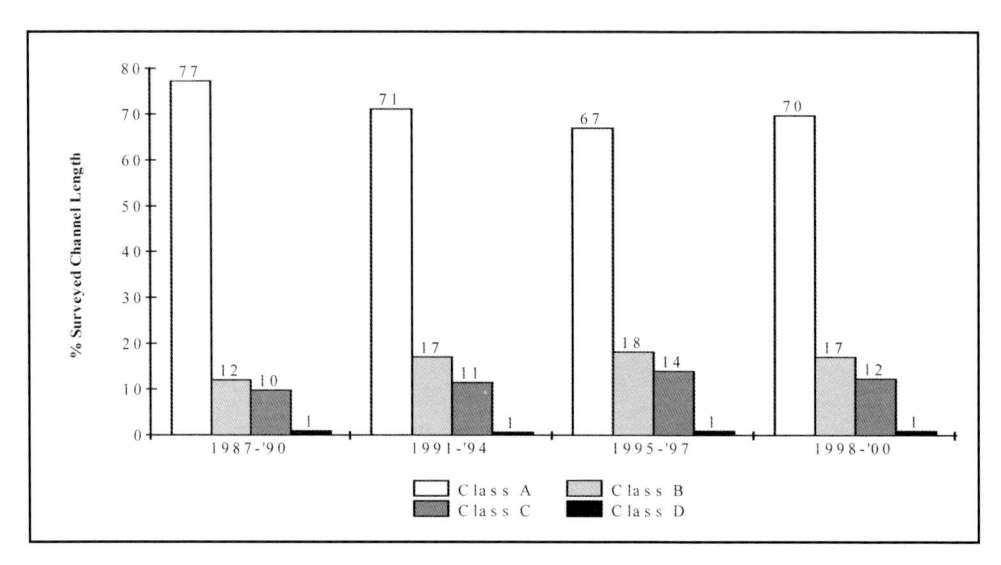

Figure 2: Recent trends in river water quality (13,200 km baseline)

from agriculture, are likely to have started to take effect in the 1998–2000 period in some catchments.

The EPA's biological monitoring programme for rivers provides an illustration of the importance of consistent and comparable long-term monitoring of the environment as a guide to the development and implementation of environmental policy.

In addition to biological monitoring, there are physico-chemical monitoring programmes of rivers undertaken by the EPA for the local authorities in the South-east, North-east and West. The most recent report on river water quality in the South-east (Neill, 2003) includes a note on river water temperatures in the region for 1979/1980 as compared to 1999/2000. The overall average river water temperature had increased from 10.52 ºC in 1979/1980 to 10.90 ºC in 1999/2000 – and increase of almost 0.4 degrees C over the two decades. The data showed that increased temperatures have occurred at the upper and lower ends of the scale of temperatures rather than in mid-range, i.e., mainly in summer and winter. It was suggested that the results might be of consequence in relation to climate change considerations.

Climate Change

A 2003 research report (Sweeney *et al.*, 2003) published by the EPA concludes that Ireland will not escape from the impacts of global climate change. The research predicts dramatic changes in rainfall patterns. Winter rainfall will increase by up to 10 per cent while summer rainfall will decrease by up to 40 per cent on parts of the south and east coasts. Other predictions include the following.

- Summers will warm by up to 2 degrees C.
- Winter temperatures will increase by up to 1.5 degrees C by mid-century.
- A sea level rise of approximately 0.5m is projected by the end of the century. In Ireland this figure will be modified by local land level changes, though a higher platform for wave attack will inevitably mean greater erosion of 'soft' coastlines, formed of glacial drift or unconsolidated materials.
- Sea level rise, coupled with suggested increases in extreme events, will render low-lying regions particularly vulnerable.
- A decline of Northern European species in Ireland can be expected, accompanied by an extension of species, including pests, that favour increased temperatures.
- Increases in some migrant insect and bird species can also be anticipated.
- Salt marshes will become more vulnerable due to a combination of climate change and sea-level rise.
- Changes in marine species distribution are projected.
- Increased sea-temperature may impact on salmon farming, diminish the commercial viability of farms and render them subject to increased algal blooms, pests and disease problems.

These and other projected changes point to the need for climate proofing of policies and plans; they also point to increased importance of long-term monitoring to track the changes. The report followed on from a study (Sweeney *et al.*, 2002) on climate change indicators for Ireland, which recommended:

> 'It is important that further use is made of the data already in existence through more in-depth analysis and integration with other data sets. There is also a need for increased co-ordination and commitment to long-term monitoring of primary (climatological) and secondary (ecological, health and economic) indicators in Ireland. In order to have an effective monitoring programme it is essential to carefully consider the choice of measurements and sampling design. These must be linked to objectives to

ensure that the results are useful, which would contribute to the development and implementation of environment policy. It is, therefore, recommended that a national strategy for environmental observations centred on the issue of climate change is devised.'

In relation to ecosystem monitoring, the study concluded: 'the value of long-term monitoring has never been more apparent. A network of long-term ecosystem monitoring sites should be established to select key indicators and collect data on various aspects of biological activity ranging from plant and animal phenology, to insect, butterfly, bat and bird behaviour.'

In Conclusion

The preparation of the new national monitoring programme has stimulated new thinking and enhanced co-ordination between regulatory agencies. It is timely because of the new international requirements referred to above. A further development is the European Marine Strategy, which is an EU initiative under its 6th Environmental Action Programme and is aimed at developing a coherent policy for the sustainable use of the seas and the conservation of marine ecosystems. It will undoubtedly be of great importance in driving monitoring activities in the future.

Overall, the marine environment is subject to a wide range of potentially deleterious pressures. Over 50 per cent of the Irish population lives within 10 km of the sea, with many of the larger towns, cities, industry and power generation located along the coast or on estuaries. In addition, river inflows bring organic matter, nutrients and other substances into estuaries from agriculture, inland towns and industry. At the same time the quality of the marine resource is extremely important to Ireland to support fisheries, aquaculture and tourism. Effective monitoring programmes, yielding policy-relevant indicators, are therefore vital to track changes in the medium and longer-term and to guide the management and protection of tidal waters.

Larry Stapleton is a director of the Irish Environmental Protection Agency and has particular responsibility for the EPA's Office of Environmental Assessment, which includes environmental monitoring, research, reporting and related fucntions. Prior to that he was a Programme Manager in the EPA and in that capacity oversaw the preparation of State of the Environment reports for Ireland in 1996 and 2000. Earlier in his career he was project manager for the Dublin Bay Water Quality Management Plan and for related projects on tidal dynamics and dispersion modelling of other tidal waters including the Shannon Estuary.

3.18
THE IMPORTANCE OF LONG-TERM MONITORING OF THE ENVIRONMENT

Michael Ludwig
National Marine Fisheries Service, Milford,
CT 06460, USA

Introduction

Humans are forever looking to the past for insights about present day and future conditions or trying to use existing conditions to foretell future circumstances. The former looks back through time to determine cause and effect relationships. The latter gathers information to manage present conditions and influence future scenarios. Interpreting past events and conditions has evolved its own sciences and practitioners, all focused on understanding the past based on the shards of evidence left by our predecessors. Every now and then the discovery or rediscovery of a diary, report or fossil provides tidbits of insight about the past. Rarely do those shards provide knowledge about the past environment. Wouldn't it be remarkable if we were able to read a text describing the environment that our forebears saw and used to survive and thrive? For instance, what was the nature of the seventeenth century ocean ecosystem before oceanic resources were harnessed to sustain human progress? What were those "target" species and how many were swimming in the sea in 1492? Early explorers of North America reported that fish were so plentiful that they impeded navigation (Squires, 1983). We know that Atlantic cod was sought in preference to all other finfish and they grew to be almost two metres long, but why and how? Possessing those insights would facilitate the management of that species and help in efforts to control human impacts on the earth's ecosystems.

Historical perspective

Looking for insights in the past leads one to wonder what future generations might find useful, for instance what is needed to establish and maintain a balanced indigenous population in the oceans? The Cobb *et al*. (2003) *Nature* article regarding the extension of the record of the El Niño Southern Oscillation and its influence on weather and fishery resources is an example of the need for historical perspective and the value of a protracted perspective for assessment of such events (Cobb *et al*, 2003). For a contentious issue example, Stephen R. Palumbi of Stanford University and Joe Roman of Harvard University recently tackled the question of the size of the historical whale populations in the North Atlantic in *Science* (Roman & Palumbi, 2003). There are no credible records of the populations. The researchers used the wide variation in whale DNA to conclude that stocks in the past were much larger than had been thought. Their finding is causing fractures in whale management programmes.

Conflicts over resources – Costs

The use and management of resources that are not owned by individuals (uplands, rivers, lakes, estuaries and oceanic) are oftentimes the source of conflicts. The 'Tragedy of the Commons' hypothesis defines these situations (Hardin, 1968). The Tragedy of the Commons is that, contrary to belief, users of commonly held resources receive little benefit from sharing or being conservative in their uses of the resource unless all parties are equally constrained. Hardin's "Commons" is a village green on which livestock are allowed to graze. Since the area is open to all, having more livestock or letting them graze the vegetation longer improves one person's livestock condition to the detriment of all others because the others are unable to obtain the same use or benefits from the limited resource (grass). In fact, it is the greedy and indifferent partners who benefit most from unrestricted access to "public" resources. Without knowledge of the size and condition of the resource (grass, fish or whales) one cannot regulate or manage that resource use to maximise the benefits (often termed "maximum sustainable yield"). To establish and maintain sustainable yields requires research and monitoring. (For the purposes of this paper surveillance and monitoring are synonymous.) Research is a well-established practice with funding problems. Monitoring has a connotation of surveillance, requires protracted sampling and has little social or economic "value" until it has accumulated sufficient information to be used. In times of limited funds, paying for products that may not be useful in the near future can be difficult to justify. Finally, funding is usually based on annual or biannual allocations. Long-term monitoring does not fare well in such funding cycles because its utility is unlikely to be realised within one or even several such cycles. These constraints handicap the acceptance, use and availability of monitoring. However, after the data establish baseline conditions, they can be the most valuable and useful source of management information. Failure to monitor environmental conditions leads society repeatedly to experience the tragedy of the commons.

Tracking changes – Wastewater impacts – Coastal zone

The use of long-term monitoring to track and assess the effect of implementing pollution remedies or the impacts of anthropogenic manipulations of the environment has become invaluable.[1] Testing the waters that receive human wastes and tracking the success of attempts to remediate degraded environments or predicting the impact of land use changes are all uses of long-term monitoring.[2] For instance, collecting sewage at centralised facilities to overcome the impacts of failed individual septic systems has become a standard practice. Unfortunately, centralised wastewater treatment has been shown to create its own set of problems. Monitoring the operation and impacts of wastewater treatment systems provides insights about those relationships as well as possible solutions to the new problems.

A good example of the utility of wastewater testing is found in the conflicts between shellfish farming, improper livestock agriculture and human wastewater management in the coastal zone. Raw Oysters, sold for consumption on the half shell, are the most valuable form of the species. Size, shape, texture, taste and colour, even the condition of the shell in which it sits, influence consumer appreciation and value. Yet it is the presence of invisible bacteria and viruses in the oyster-growing area that most often regulate harvesting and sales. This sensitivity to water quality makes shellfish farming a sentinel for water pollution. Because of their sensitivity, sentinel species health, abundance or presence can be used as warnings of change in the area and that action may be needed or worse conditions can follow. Pollution from point and non-point source waste discharges (respectively, pipe discharges and runoff from open areas) released into coastal waters can degrade those waters and force the termination of aquatic farming activities. More development means more potential pollution opportunities with more potential for water quality degradation. Likewise, inadequate waste management on a farm produces similar situations. Excessive use or poor application of fertilizer, pesticides or inappropriate animal waste management practices can result in immediate or sporadic releases into local waterways. The moving force might be direct input of the materials or it might be facilitated by rain induced runoff. In either case, the movement of materials from the land to the water is a constant event: adding substances to it does little for the receiving waters. Foregoing shellfish farming is not a practical option for several reasons, not the least of which is job and income/crop loss. More importantly, loss of shellfish farming due to pathogenic pollution portends loss of human health protection during other water-contact activities. Loss of the ability to use the water safely, eliminates many of the very reasons that people are locating near open water and can destroy tourism as well as reduce real-estate values. Monitoring water quality and using the data to provide early warning that problems are developing and modifications are needed serves two purposes: it continues to allow the regulation of shellfish farming and it provides a warning that changes are needed in waste management policy and practices.

Coastal waters belong to the people of the abutting coastal state. It is a responsibility

---

[1] See Global Terresterial Observing System (GTOS) at < http://www.wsl.ch/forest/risks/wsidb/gtos/gtos.html >
[2] Retrieved from U.S. EPA at < http://www.em.doe.gov/techguide/techguide.pdf >.

of their resource trustees (Government) to insure that water quality is not impaired. Although the relationship between pollution and consumption of safe shellfish is fully appreciated and the protection of water quality is not within their direct influence, it is seafood harvesters, often, that suffer the consequences of reactive rather than proactive recognition that water quality is being degraded. All too frequently, rather than terminating the source of the pollution, the recipients of the degraded water quality are penalised through closure of shellfish growing areas. Pre-emptive practices that protect water quality are routinely less expensive undertakings than waiting for a crisis before reacting. Typically, and sadly, it is only after a crisis occurs that corrective measures are considered. This mindset was changed in the U.S. in 1972 with the passage of the Federal Clean Water Act. The objective of that legislation is to restore and maintain the chemical, physical and biological integrity of the Nation's waters. The European Union took a similar step with the passage of the Water Framework Directive (European Council, 2000). The goal is to attain and maintain water quality that is "swimmable and fishable."[3] A well-considered long-term monitoring programme is the tool that can provide early warning as well as opportunities for pre-emptive action, if it is implemented correctly. All too often, water quality monitoring is sporadic or insufficient to the tasks assign it because the objectives of monitoring are not fully characterised or implemented.

Sorting out natural from man-made variations – The importance of data sets

The earth is in dynamic equilibrium and change is a constant component of that condition. Sorting those cycles and patterns from long-term change is a challenge, as is balancing human use of the environment with the obligation of sustaining renewable resources. Both tasks require appreciation of natural variations and cycles. Whether looking locally at water quality conditions, or globally at the movements and habitat use of highly migratory species such as whales or Atlantic blue-fin tuna, the collection and publication of long-term monitoring information has untold value to decision-makers charged with guiding the use of those "resources". All too frequently, data collection practices remain disjointed and incomplete, providing little more than snapshots of conditions at one or a modest few locations and moments in time. These snapshots suffer from a lack of the substance and breadth needed to define natural variability and so, while appearing to increase the reader's confidence in an action they often misrepresent the facts and increase the uncertainty of the reviewer, regardless of the precision applied during data collection. Long-term monitoring provides a broader and more comprehensive view of conditions and events by containing the texture of natural variability within a data set.

Casual but long-term observations of the date on which perennials flower in backyard gardens and migratory birds appear at locations along their migratory routes have been used to characterise the effects of global warming. These anecdotal observations

---

[3]Federal Water Pollution Control Act Amendments of 1972. 33 USC, 1251 et seq. viewed at
< http://www.epa.gov/region5/water/pdf/ecwa.pdf >

provide evidence of change by characterising the occurrences of usual, unusual or unexpected appearance of species or changes in their presence. This helps us appreciate the consequences of change while revealing that spring arrives about two weeks earlier than just a few decades ago. Fishermen are reporting changes in catch composition. And, the impact of the changes is complicating fishery management efforts at stock recovery. How can we use knowledge of these events to manage our biosphere? The answers are not always clear but the value of the monitoring information is limited, often, only by its scope. It is apparent that without data to characterise the changes, we have no basis to make comparisons. The United Kingdom has a programme that provides its citizens with the ability to provide such information.[4]

The Mid-North Atlantic is a boreal ecological zone possessing a temperate climate. Many of the species that use this ecological region are migratory, moving south in winter and north in summer. Species using the area travel from both the north and south, depending on the season of the year. For instance, it is common to see seals and sea lions "summering" in the waters off Rhode Island, Connecticut and New York from November through April and sea turtles "summering" from May through October in the same areas. This flux of species indicates that different organisms are using the same habitats but at different times. How and why species use the region and the functions and values afforded them by the habitats within the region are more difficult to interpret than the functions and values afforded by more stable regions with resident species using habitats such as coral reefs. These differences make understanding potential impacts and developing decisions affecting the environment more difficult to undertake without a comprehensive information base.

Environmental data for decision-making

With the realisation that human choices can have significant environmental consequences, comes the recognition that comprehensive data collection and its assessment are vital tools for resource managers charged with decision-making. The use of long-term monitoring to expand and improve understanding acts to help reduce uncertainty and the risk of mistakes – and the undesirable consequences that mistakes allow. However, postponing development activities until comprehensive impact evaluations can be developed is impractical, occasionally. To overcome the unavailability of site-specific information studies, often times, brief visits or limited sampling efforts at a site are used to characterise it. Brief and incomplete evaluations of a site are often termed 'snapshots' because they are based on a view that occurred only that day but doesn't have additional supporting information to determine if the picture is complete. Snapshot environmental characterisations may actually capture the seasonal presence of a species and it is possible, but unlikely, their maximum abundance as well. It is more likely that these evaluations will undervalue the importance of a site. The undervaluing results because they routinely miss the peak use period(s), they can't capture the seasonal and annual variations of a resource use or habitat relationships and the sampling or evaluation may not even approximate 'normal' conditions.

---

[4]See < http://www.woodland-trust.org.uk/phenology/newsletter.htm >

What is 'normal'?

As an aside, "normal conditions" seem generally unavailable as our global climate reacts to induced change. For instance, how might one characterise the utility of data collected around the North Atlantic during the fall, winter, spring and summer of 2002–03? The period will be remembered for frequent and intense storms, initially persistent (but not intense) cold followed by record heat, all complicated by unusual amounts of precipitation. These are unusual events and the environment appears to have responded in unusual ways. Preliminary assessment of spawning shellfish and finfish reveals they were remarkably unsuccessful in this region during the spring and summer of 2003. Is this an anomaly or portend of the future? Do data collected to characterise the importance of an area get the same weight this year as information collected in prior years? Might the study have "missed" finding an area that had great importance? Long-term monitoring can answer those questions. The most obvious use of long-term monitoring is to provide averages of use and ecological value. What is the "normal" use level? It is somewhere between zero and something as expressed by the range of values used in establishing the average.

Starting points – Insights

Long-term monitoring, in general, starts with a need for insight. It tends to be a general questioning of large or landscape-scale topics but can evolve or give rise to specific focus investigations. For instance, what is the seasonal composition of the phytoplankton community in the Bay of Fundy or what is the impact of discharging sewage effluent into a waterway? The initial intent of a long-term monitoring effort might be to identify and characterise the dominant plankton species and their patterns of presence along with their location in some water body. Weekly plankton sampling performed throughout the year or during peak growing conditions would capture the presence and absence patterns and, over time, more and more of the species variations. The information collected could be used, initially, to site shellfish aquaculture facilities in the most advantageous conditions or reduce the likelihood of growing crops in an unhealthy, nuisance species (red tide) hotspot. Finding the place where red tides are routine occurrences could call for another monitoring of that specific location. In the case of sewage treatment facilities, monitoring might be used to insure continuing compliance with requirements of a permit but provide the additional benefit of depicting the effects of local development by tracking the increases in discharges. The data become a verification that the plant is meeting non-degradation objectives although development of an area is occurring. A case in point is the Boston offshore discharge of their treated sewage. In the late 1990s, Boston began using a 15.3-km sewage outfall pipe. Five years after starting the discharge, the monitoring programme may be revealing unanticipated consequences of the activity. Sampling, in the warmer months, when sewage discharges increase and oxygen is naturally in short supply, might be the only period when data are needed. The monitoring should provide situation verification, a more comprehensive understanding of cause and effect, insights on impact variability and serve to refine the impact expectation. Boston's sewage may require further processing to regain the protections sought by its designers.

For more than sixty years, the City of New York used dissolved oxygen levels collected at forty stations in New York Harbour to characterise the impact of their wastewater discharges. The assumption was that oxygen depletion, due to bacterial decomposition of the sewage released into the Harbour, was the cause of their degraded water quality. It appears, now, that this hypothesis was overly simplistic.[5] Holding and decomposing sewage within the plants (processing) increases the availability of nutrients in the discharge and facilitates algal blooms near the discharge point. Managers of the plants now recognise that the characteristics and volume of the discharge are vital elements for effective pollution control. Without their long-term monitoring programme, this relationship would not have been fully appreciated in such a timely manner.

How are successful monitoring programmes started? There is no single answer to that question, but a start lies in the purpose to which the information will be put and occasionally, with serendipity. Weather data collection was as much serendipitous and laziness as good and methodological science. The initial and principal efforts of weather data collection were related to weather prediction for the immediate future. People collected information and didn't throw it away for a day, week, month or year. Those data accumulated until someone could compare year-to-year events across decades and discover patterns and cycles of events. Applying knowledge about weather patterns helps reduce the risk to farming, travel, construction and other activities by reducing the likelihood of failure and lost investments. Curiously, for the 2002–03 winter, the most accurate long-range weather predictions were found in the *Farmers Almanac*. Those predictions are based solely on interpretation of old weather records![6] Because so many human activities depend on weather conditions, we routinely collect the data from many locations. The US Weather Service is collecting its 110th year of comprehensive weather data in 2003. They use "X" number of measuring stations as well as satellites to develop their predictions. We have weather records but don't ask how large the flounder populations were before we started harvesting them or how finfish populations in general, reacted to natural variability. Those insights are usually anecdotal, limited in scope or simply, unavailable, but the existing tidbits of information we do have are tantalising.

Confidence – Duration and breadth of a data record

The length and breadth of a long-term monitoring record needed before it can be called upon with confidence is not standardised but is dependent on the natural variability of the conditions in which an event or condition is monitored. Ultimately, the magnitude of necessary monitoring activities is directly dependent on the nature and magnitude of the uncertainties in the management decision. Simply, the more data one collects the lower the likelihood of choosing the wrong solution. The weather record for 2003 may become a single anomalous piece of the larger data set of annual weather or

---

[5] Retrieved from the World wide web site of the U.S. Environmental Protection agency (EPA) at:
< http://www.epa.gov/owow/tmdl/ >.
[6] Farmers Almanac. 2003. Press release. Retrieved from the Farmers almanac at < http://www.almanac.com/ >

it might mark the start of a change in conditions effecting our environment for generations to come. Fortunately, we have it and the larger record to revisit in future years. Long-term monitoring data collection provides those opportunities. The seasons change in annual patterns but the lowly Bay Scallop lives less than two years and has only one spawning period in which to sustain its species. Should they fail to reproduce or even fail to attain moderate reproductive success, the local population faces extinction. Determining what occurred during and after a spawning event requires one brief sampling period. To understand the variability of reproductive success or failure in the context of the natural forces at work during the spawning period requires monitoring of multiple spawning events. These are different questions with different answers. However, a single day of sampling within the Bay Scallop's month long spawning period may prove as uninformative as the single year of weather sampling. Clearly, the length of the monitoring is tied to the parameters of the event, the degree of natural variability and the objectives of the investigator. Reliance on data from a short-term collection effort or a data set that is insensitive to the patterns of the target species, regardless of sampling duration, may produce inaccurate insights and unacceptable standards.

Conversely, taking the air temperature at some chance location on odd Wednesdays will not provide information that might foretell the future unless the locations are nearby and the data are collected for a prolonged period. Variability of data within a record controls the confidence that can be placed on it and usually requires the use of statistics to create the confidence ranges or help determine the likelihood that the data are accurate. Managing the risk of error and hypothesis-failure has been enhanced by the mathematics of statistics. However, uncertainty is risk and statistics do not eliminate risk. In fact, statistics may reduce confidence in a record by their inability to explain deviations and complicate rather than clarify issues. In the end, reducing uncertainty requires building a useful record of the items of interest in appropriate formats at meaningful intervals. Unfortunately, comprehensive monitoring programmes are not routinely established, maintained or administered in such a way that those criteria are met. Curiously, so much of what is used to manage present resource allocations or predict future levels of availability is based on the record of past events and faulty and incomplete data. Invariably those handicaps cause both environmental and economic hardship. A well-designed long-term monitoring programme can preclude or resolve those problems.

Models as an aid to decision making

To improve the ability of decision-makers to select correct solutions, modelling is often used. Models use real data to create artificial representations of the parameters under varying conditions to depict possible outcomes or variations of an event. Expediting evaluation, reducing delay, providing a visualisation of events or consequences along with the degree of confidence in the likelihood of each option occurring and purporting to limit unexpected complications are all benefits ascribed to models. At the heart of any model is the set of existing conditions, collected in the field, and a series of assumptions defining how or why those conditions occur. To improve a model's accuracy requires increasing the knowledge regarding the variables and alternatives upon which the model is built and operated and defining the possible outcomes with better representations. To achieve higher degrees of confidence, modellers are constantly moving toward more complex and dynamic model structures that require more accurate or complete information.

Model outputs often imply that they can determine the best alternative. Because assumptions are used to create the model, incomplete understanding or incorrect use of the variability within those assumptions can fatally flaw the outcomes. If the consequences are incorrect any decisions are faulty as well. The best solution to these problems is more data collection and for a longer period of time. As with direct use of long-term monitoring data sets, modellers can accept some error or data gaps within a large or long record and continue to obtain the designers objectives. In situations where the data are unusual, they sit outside the trends and patterns and might be dealt with by evaluators. Models that rely on short-term collections of data do not possess the same luxury. Because the challenge in the environmental management business is to authorise only those activities that are environmental friendly or passive in nature, long-term monitoring data sets are a sought-after commodity.

The challenge faced by managers of renewable resources is sustaining those resources at acceptable levels in uncertain times. How might one know when renewable resources are at a balanced indigenous population level? The answer is, we frequently don't know, and management guesses (occasionally conjured up) reflect this condition as well as the limitations engendered by the available information (Quinn & Deriso, 1999). Fishery management agencies use data, statistics and models to determine what is present and how it might change with time. Looking at the success rate of decisions based on limited information regarding aquatic resources reveals a disturbing trend; the decisions are incorrect far too frequently. For the Canadian's Western Atlantic cod stocks, the data were insufficient to the challenges and the interpretations were overly optimistic.[7] That cod fishery collapsed from overfishing. The models used by Canada and others have shown that we haven't collected sufficient information about natural variability, its effects on individual stocks or the resource community at large to successfully predict future conditions. As model results continue to be at odds with perceptions, unable to fully characterise the range of variability in situations or unaccepted by the community, there is a recognition that physical monitoring must be more comprehensive or renewable resources will not attain or maintain sustainable levels.

Importance of understanding natural variability

Can these problems be overcome? In many cases the answer is "yes." Understanding natural variability is critical to correcting the problems. Conditions change and that change and its degree of persistence are indicators. Identifying the changes or their indicators is the first and most critical component of the process. How the change might occur or what the change might cause must be determined. While an answer may be

---

[7] MacKenzie. D. 2002. The Downturn of the Atlantic Cod (*Gadus morhua*) in Eastern Canada; What is happening to these fish, and why? Retrieved from World Wide Web site < http://www.fisherycrisis.com/nscod.htm >.

divined from a basic monitoring programme, it often takes a more comprehensive assessment of conditions to find the complete answer. For this reason many monitoring programmes incorporate a tiered or stepped approach which has proven to be the most "results-oriented," cost-efficient and objective method of collecting data. The specific design of a tiered monitoring programme depends on the objectives but, typically, the approach starts with monitoring of physical attributes. In oceanic situations, the physical tier includes oceanography, geology and coastal processes. When deviation from the expected outcome of the monitoring occurs, the effort can be enhanced to include assessment of more physical items or more chemical parameters such as water or sediment, chemical analyses or characterisations which are useful for evaluating pollution problems. The most difficult information to collect and interpret but the most comprehensive are biological characterisations and they are usually the highest or last step of a tiered testing programme. In some monitoring efforts, it may be possible to have a tiered approach within a single category. The management of dredged material relies almost entirely on sediment chemistry profiles and pollutant availability to characterise the suitability of placing dredged material in offshore disposal sites. Once the causative factor(s) of an unacceptable change are identified, it may be possible to alter them or develop measures and practices that accommodate the change. For example, habitat loss is routinely identified as one of the main causes of changes in community structure. Collecting data on what constitutes a "good" habitat creates the basis for determining what steps or measures might be invoked to recover from, remediate or redress undesirable conditions. In the last two decades, the practice of creating habitat has moved from "attempts" to "expected success" with salt marshes and shellfish, lobster and finfish habitats.

Managing rainfall

Building houses, roads and car parks reduces the amount of area where water can be absorbed into the ground. That would be enough of a cause for concern, but when rain falls on those impermeable structures, adjacent areas must absorb the added volume or pass it along. To address this problem, developers usually offer water management measures such as drains and swales that direct water into local waterways. The increasing volumes of water moving into drainage systems created by each additional development increase the potential for erosion. The use of devices for rapid relocation of the water (such as stormwater drains) can induce flash-flooding downstream. All these consequences can destroy habitat within natural waterways. However, there are unseen consequences of incorrect management of precipitation at development sites as well. If the precipitation occurs in summer, the water flowing off the structures may be unnaturally hot and incapable of holding necessary amounts of oxygen or it may simply exceed the range in which survival of the resource is possible. However, when a development is proposed or water quality monitoring is developed how might these events and conditions be captured so that resource management and land use planning are effective? Often times they are not and we lose the resources within the waterway. The trend has become so obvious that major rivers are causing flash-flooding and this is occurring after rainfall amounts once considered to be only moderate in volume.

Resource Managers and Planners need to know what might happen if an activity is undertaken. How that information is gathered, presented and addressed are challenges. A single or even group of samples taken during a day, week or month probably fall short of adequately characterising the variability of a stream flow. Yet it takes only one excedence of a limiting parameter for the briefest period to impair the aquatic resources in a stream. Either targeting the periods when an event might occur or shaping a long-term monitoring programme to collect sufficient data to overcome the variables and vagaries is needed. The former option requires high flexibility and mobility of the field team; the latter requires protracted and persistent sampling, possibly obtained by data recorders set in place for days or weeks. A well-designed long-term monitoring programme is the better choice and can hit the target objectives and others at the same time. For instance, determining what physical, chemical and biological conditions are needed for spawning Atlantic salmon can help set standards for runoff- and point-source discharges associated with development activities.

Amateur and professional data collectors – Engaging the public

Who can become a long-term monitoring data collector? While degrees from advanced learning institutions facilitate development of the monitoring, the collection activities need not be as constrained. The types of data collection range from simple temperature of an environment to complex measurements of atmospheric ozone and water quality, with the list growing every day. To staff these growing efforts is a body of individuals ranging from professionals to amateurs. The professionals bring focused skills; the amateurs often bring just the desire to help and the hope that their efforts will advance understanding. In fact a cottage industry has been established connecting volunteers, willing to pay their own way, and researchers in need of assistance. These groups provide evidence that it takes only three attributes to be good data collectors:

    a) willingness to possess the skills needed to perform the sampling methodologies,

    b) the willingness to partake in the activity and

    c) the drive to collect the information.

The papers presented in this book were made possible by a wide diversity of individuals with those three attributes.

In the last decade, many nations have seen a major increase in the use of residents to collect information on all manners of topics. Many of these people are volunteers, collecting information in long-term monitoring programmes that are both ongoing and just beginning. Structured to the level of data collection needed and sophistication of the samplers, a modest number of long-term monitoring records are being established for a plethora of topics. Using field staff comfortable with the sampling technique increases confidence that the information has validity. Project W.E.T. (Water Education for Teachers) helps school children collect and disseminate water quality information as a part of more comprehensive learning programs.[8] Resource managers train teachers in

---

[8] See Project WET (Water Education for Teachers) at < http://www.montana.edu/wwwwet/ >

the use of measuring devices and expected results (ranges) to a degree where use and benefit of the student activities is possible.[9] Recreational boaters are provided sampling and tagging directions which allows them to collect statistics on their catch and helps track released species.[10]

Reaching out and engaging the public may be the hardest task in setting up a monitoring programme. Fortunately, there are programmes and professionals available to help in those efforts. Once volunteers are engaged in activities that meld with personal goals or enjoyments, the process becomes one of coordination rather than enlistment. The Non-Governmental organisations (NGO) *River Keeper* and *Harbor Watch* are activist organisations that monitor the health and welfare of waterbodies. The *River Keeper* programme began in the Hudson River but chapters of the group can be found monitoring activities affecting waterways from New York Harbour to San Francisco Bay. *Harbor Watch* is a local group that focuses on the health of harbours in the western portion of Long Island Sound. The two NGOs operate monitoring programmes that collect water quality and finfish information.[11] Training members in sampling methodologies and data management was initiated by their leadership then augmented by State and Federal regulatory agencies. The groups use traps, trawls and seines to collect finfish and other mobile species and water quality measurements to characterise waterway health. The water quality of western Long Island Sound is sampled with sufficient frequency by members of these NGO groups that potential hypoxia (low dissolved oxygen) events are now predictable days before their actual onset. The data are passed to resource managers within hours of collection. The work of these groups has proven to carry some unforeseen benefits such as providing insights on permit compliance and tracking the results of sewage treatment plant upgrades. Another programme that has found good success using trained volunteers is the U.S. Fish & Wildlife Service's migrating shoreline bird census. That programme is almost completely staffed at the field level by trained volunteers.[12]

The valuable data set

Example – Salt Marsh Plants

What makes a comprehensive and useable long-term data set? The answer to this question appears to depend on the nature and frequency of occurrence of the range of variables under examination. Starting with a single sentinel of conditions can be the best technique for characterising variable settings. For instance, the plants in a salt marsh are sensitive to a number of conditions. Those species in the "high" marsh zone (only occasionally inundated) are particularly sensitive to the degree of inundation. Salt marsh plants in the northeast U.S. grow and reproduce in about seven months. The plants start growing in spring, mature in summer and flower in late summer into early fall, dispersing their seeds during the fall and winter.

---

[9]For example see < http://www.outreach.washington.edu/k12guide/resourcepage.asp?ProjID=248 >
[10]See the American Littoral Society webpage regarding recreational fishers tagging programme < http://www.alsnyc.org/ >
[11]For more information, see < http://www.riverkeeper.org/ >
[12]For more information see < http://migratorybirds.fws.gov/shrbird/shrbird.html >

One could study the inundation of the plants for the seven "growth" months and, barring catastrophe, be able to adequately characterise the "needs" of the plants and the some of the functions and values being provided by their tidal habitat. Conversely, variability of the chemical environment and its impact on the growth of the plant could be the objectives of monitoring. With the knowledge of those needs, changes in plant condition would be recognised and, if necessary more intense scrutiny of the site conditions could be undertaken. The health and productivity of those plants or a specific species within the group could provide insights about sea level rise, nutrient availability, pollution, waves and climatic change. The plant's success or displacement by other species could be used to indicate that change is occurring. The monitoring information could be used to guide plantings at other locations. During the non-growing season, portions of the plant may remain on-site or be dispersed and this period of non-activity may not require or merit monitoring. Looking at one species, salt marsh cordgrass (*Spartina alterniflora*) is enlightening. The species grows between mean sea level and the high tide line. It is revered and well studied along the Atlantic Coast of the U.S. but considered an invasive, nuisance species and threat to unvegetated intertidal flat communities and shellfish farming in the Pacific Northwest. Knowledge garnered along the East Coast is being used to restrict the proliferation of salt marsh cordgrass on the West Coast and insuring that the natural conditions are not diminished.[13]

Example – Spawning Fish

The study of the spawning needs of a migratory finfish could be quite limited, provided the fish show up to spawn. (Failure to appear could indicate that the access corridor no longer provides the functions and values necessary for the migration to occur. Such an alteration of habitat between the ocean and riverine spawning sites is a major component of the problems facing Pacific Coast salmon stocks.) Monitoring a finfish spawning activity might require but a few days of focused attention at a spawning site. Repeated, annual collections of data (long-term monitoring) for the same amount of time would accumulate information on the variables, much like the bay scallop investigation discussed above. Determining spawning success and the influence of physical forces might be the initial objectives. The selected sampling parameters could include habitat availability and a juvenile index of survival (sampling the number of young to describe the spawning success). Although the actual spawning event could be understood in a modest period of time, quicker if supplemental work were done in a laboratory where variables can be controlled, the variations of population cycles and habitat quality are best seen in a prolonged field data set. The long-term monitoring sampling could be undertaken by school-aged individuals with support from local experts. The products could be used to control development impacts and set assimilative capacity standards that allow for some environmental deterioration of the waterway before remediation or use restrictions would be necessary.[14] However, waiting until one or a series of unacceptable events has occurred and then relying on forensic science to develop an appreciation of the

---

[13] 2000 Spartina Management Plan for South Puget retrieved from < http://www.willapabay.org/~coastal/nospartina/
[14] See US EPA Total Maximum Daily Loading (TMDL) at < http://www.epa.gov/owow/tmdl/ >

pre-impact condition is problematic. For instance, the relationships between Atlantic salmon and their riverine spawning site habitat remains poorly defined.[15] Neither the ecosystem that supported the natural populations of those species nor the variations in species presence was ever characterised to a degree that allows understanding of the species or their place within the ecosystem. Those shortcomings may account for the uncertainty and problems associated with recovering stock sizes, community structure and the methods to attain those objectives.

Becoming a data collector

How can one become a long-term monitoring data collector? The first challenge is to find a programme that is collecting information. The characterisation of Urban Forests (City trees) is a federally funded monitoring programme in the United States used to determine the characteristics and health of publicly owned forest stock in urban and suburban settings. The programme requires walking the public roadways of a community, collecting the statistical facts regarding setting, health and dimensions of publicly owned trees along the roads and recording it in a useable formats such as a Geographic Information System (GIS). In one instance, the staff for that programme came from the ranks of walkers, hikers, photographers, and other people willing to offer their services for the task.[16] Free training in tree identification was used to attract participants and arrangements made at the initial contact for subsequent meetings. Note that prior knowledge of tree species identification or skills with data collection devices was not a prerequisite. In fact, the offer of free training was the device used to establish the volunteer corps. Training in a group or on-the-job setting appears sufficient for many types of monitoring. The Urban forestry group collected information on more that 17,000 trees over a two-year period. Today, their community has a maintained inventory of its trees that can be referenced to manage and maintain that resource. The incorporation of the data into a City's GIS provides mapping and other site information helpful to City Planners, Engineers and Arborists.

Quality requirements

What is the quality of data needed? As noted above, the quality of the data must fit the need or challenge of the situation and possess appropriate accuracy. Members of the tree inventory group identified representatives of more than fifty tree species with an error rate of less than one percent. Conversely, there have been instances when species identification has required a high degree of expertise and significant errors were found in their work. The errors diminished, initially, the validity and value of data sets, but re-examination of the samples was able to resolve the conflicts and improve the data quality. To overcome identification problems, reference collections of species or guides to their identification (keys) have proven to be invaluable for all types of sampling. Again, it appears that the successful melding of staff and methodology are critical keys to success.

---

[15] For example see < http://archive.greenpeace.org/comms/cbio/cancod.html >
[16] See Milford, CT Urban Forestry volunteer programme(Leaflets) at < www.norwalktreealliance.org >.

One benefit of collecting data for long periods is that an occasional, anomalous data point does not diminish it. How many whales are there in the ocean? Finding a method of accurately counting them is the first challenge. The discovery that each whale has a unique colouration pattern and shape to its tail flukes made tracking and enumerating possible. Confidence in the data set can be assured with a reasonably good photographic recording of the fluke pattern. The more pictures taken over time, the better the confidence in the identification and the more robust the data set.

Similarly, dissolved oxygen (DO) levels are one of the most commonly used measures of the environmental conditions of waterways.[17] Water holds only so much dissolved oxygen. Water temperature as well as the rate of use and replenishment controls the amount. Along with the amount of light penetration (often measured with a Secchi disc), DO levels are the preferred, first tier, measure of water quality. The two measurements provide many of the insights that resource managers need. Improvements in sensing devices over the last two decades have enabled the simple, quick and accurate depiction of those parameters. Accuracy of DO measuring systems is in tenths of a milligram per litre (mg/l) and light penetration is accurate to less than 30 cm. In coastal waters, the DO level ranges between 0.0mg/l and approximately 14mg/l and light penetration depths range from less than one centimetre to over one hundred metres. Both recording limits are more than sufficient for good management decisions. In the end, sampling precision must be set by the needs of the situation.

Influencing the decision-maker

Recognition of the importance of long-term monitoring and the commitment to undertake it expressed by people who are responsible for funding decisions remain problematic areas. One solution to the impasse is to incorporate long-term monitoring in permits and authorisations and rely on resource users to gather the evidence of the consequences of their actions. Those allocated permits are, after all, using public resources for private gain. The monitoring of discharges to waters in the U.S. falls under the National Pollution Discharge Elimination System (NPDES) programme of the Clean Water Act.[18] Permit recipients must collect and report compliance with their permit conditions as well as any deviations, or suffer the consequences. Clean-up of hazardous-waste sites are tied to long-term monitoring activities to set timelines and determine the level of success of a clean-up. Another monitoring initiation trigger can be seen with events regarding a resource that engenders public concern. In these cases, the public pressure facilitates the financial commitment. Whales, dolphins, the highly sought-after striped bass, as well as the species of fish harvested for human consumption have all become the recipients of long-term monitoring programmes as the result of public pressure to insure the wellbeing or continued presence of those renewable resources.

The use of juvenile indices of abundance to characterise the future size of a stock is a common practice that has produced many insights and management decisions. For

---

[17] Interstate Environmental Commission. 2002 Annual report. Retrieved from < http://www.iec-nynjct.org/reports.htm >
[18] Information on the Clean Water Act NPDES Program retrieved from < http://cfpub.epa.gov/npdes/ >

example, striped bass spawn in relatively few estuaries along the East Coast of North America. Annual, late summer and early fall monitoring of the number of young-of-the-year in and around their natal estuaries can provide information on reproduction effort and success. The index provides, as well, the number of potential adults entering the population. Collecting the annual reports creates a long-term monitoring record of the population. Juvenile indexes were used to justify closing fisheries and later, to set catch quotas and size limits when the stocks began to recover. The long-term monitoring continues and each year provides managers with information on population size and future availability.

Who benefits?

Finally, who benefits from the collection of long-term monitoring data sets? There is no complete or precise response to this question. The answer depends on the nature and character of the data being collected and to some extent, the objectives of the long-term monitoring designers. But, it is possible to say, generically, that we all benefit from long-term monitoring and the uses to which it is put because the information helps us understand our ecosystem and regulate the actions that might alter it. Today, resource managers are required to evaluate and resolve environmental conflicts. The insights needed to make those choices and manage the nature and extent of induced impacts can be found in long-term monitoring data. When interpreted and placed in an appropriate context, the information tells a story or may set the stage for further investigations. Impacts can be avoided, minimised, mitigated or compensated for, but it is necessary to know, beforehand, what the conditions are and the possible variations that might occur as the result of the action.

Environmental Assessments; Environmental Impact Studies

One of the newer resource management tools that much of the world has come to rely on is the environmental assessment. Those considerations and documents benefit from long-term monitoring information by enabling a more comprehensive understanding of the consequences of proposed actions. In their most exhaustive form, environmental assessments become environmental impact statements (EIS). The premise of drafting an EIS is that there are environmental impacts associated with the action. The EIS allows the proponents or regulatory agency the opportunity to describe the existing environmental conditions, the proposal and ways in which they plan to avoid, minimise or mitigate adverse impacts or explain why the social value of their actions exceeds its environmental consequences: (for instance, building a sewage treatment plant in a salt marsh). How can decision-makers be sure that the criteria regarding contents have been met if the document is based on information collected during one or two field trips to the site or experiences recorded during the winter of 2002–03? Drafting an accurate assessment document is difficult. It needs comprehensive information on resources and risk as its foundation (existing conditions). And, although an EIS should be written for the general public to understand, people familiar with the resources and situations discussed in EIA documents evaluate them best. As an example, the EIS documents for fisheries are complex and difficult to read unless the reviewer understands the practices.

Unfortunately, and all too frequently, the basis of an environmental assessment document is an incomplete or limited characterisation of the existing conditions. Use of information that recognises only a portion of the variability of events, habitat functions and values of the resources found within the impact area precludes comprehensive evaluation and resolution of adverse impacts. The requirements for developing accurate and comprehensive environmental assessment documents were drafted to assure that planning and authorising agencies be fully appreciative of the consequences of applicant's and their own actions. When done without an adequate database, the public and regulatory reviewers, who must determine the adequacy of the evaluation and appropriateness of an action, are handicapped. Environmental and economic "costs" of incorrect resource management decisions become public burdens. We pay for failed septic systems or faulty drainage systems by expending tax revenues on sewage treatment plants and flood control. The value of long-term monitoring information in an environmental assessment takes many forms. Its use in "completeness" requirement becomes more obvious with each decision regarding the management of natural resources. How thoroughly we embrace the concept and practice of long-term monitoring and the outcomes derived from devoting more time to recording environmental conditions will influence, greatly, what future generations will find when they look at the resources we managed for them. And, the information collected today will be available to help guide future generation's decisions.

Michael Ludwig has been employed by NOAA's National Marine Fisheries Service (NMFS) for more than twenty-five years in the Habitat Conservation Division. His principal responsibilities are related to evaluating the environmental impacts of development with Coastal and Exclusive Economic Zone waters of the United States. He is Biologist in-charge of the Milford, Connecticut field Office and the NMFS, Northeast Region's Aquaculture Coordinator for the Division. His education includes degrees in Zoology and Marine Geology, Marine Science focusing on predator prey relationships and Physical Oceanography.
Mr. Ludwig is involved in national and international resource management issues such as subaqueous utility installations, the use of transgenic species, water quality enhancement and dredging. Specific activities include dredged material management, use of explosives in aquatic settings, habitat enhancement, shellfish stock restoration, the evaluation and identification of managed fishery species essential habitat needs, endangered species management as well as the assessment of Navigation improvement projects for the principal Ports in Connecticut, New York and Rhode Island.

REFERENCES

Atlantic States Marine Fisheries Commission. (2002) 2002 Stock Assessment Report for Atlantic Striped Bass. 36th SARC. Draft Advisory Report. 72 pp.

ANGEL, J.R., D.L. BURKE, R.N. O'BOYLE, F.G. PEACOCK, M. SINCLAIR, and K.T.C. ZWANENBURG. (1994) Report of the Workshop on Scotia-Fundy Groundfish Management from 1977 to 1993. *Canadian Technical Report of Fisheries and Aquatic Sciences.* 1979 vi + 175 pp.

BAKER, J.M., HISCOCK, S., HISCOCK, K., LEVELL, D., BISHOP, G., PRECIOUS, M.,COLLINSON, R., KINGSBURY, R. & O'SULLVAN, A.J. (1981) The rocky shore biology of Bantry Bay: a re-survey. *Irish Fisheries Investigations. Series B,* **23:** 3–30.

BALECH, E. (1995) *The genus Alexandrium Halim (Dinoflagellata).* Sherkin Island Marine Station, pp151.

BARNTHOUSE, L. W., KLAUDA, R. J. & VAUGH, D. S. (1988) Science, Law and Hudson River Power Plants. A Case Study in Environmental Assessment. Introduction to the Monograph. *American Fisheries Society*, Monograph **4:** 1–8.

BERLAND, B. (1993) Salmon lice on wild salmon (*Salmo salar* L.) in western Norway. In: Boxshall, G. A. & Defaye, D. (eds), *Pathogens of wild and farmed fish: sea lice,* pp. 179–187. Ellis Horwood Ltd., West Sussex, United Kingdom.

BISHOP, G. (1998) Twenty Years of Baseline Studies at Sherkin Island Marine Station – What they tell us. *Proceedings of the 14th Annual Environmental Conference, Sherkin Island Marine Station,* pp. 33–39.

BISHOP, G. (2003) The Ecology of the Rocky Shores of Sherkin Island – A twenty Year Perspective. Pub. *Sherkin Island Marine Station*, 305pp.

BOLGER-HYNES, E. (1999) A Five Year study from 1995 to 1999 of the Flora and Fauna of the Rocky Seashore of Cork Harbour. Sherkin Island Marine Station (Unpublished report).

BRANDAL, P. O., EGIDIUS, E. & ROMSLO, I. (1976) Host blood: A major food component for the parasitic copepod *Lepeophtheirus salmonis* Kröyer, 1838 (Crustacea: Caligidae) *Norwegian Journal of Zoology*, **24:** 341–343.

BRISTOW, G. A. & BERLAND, B. (1991) A report on some metazoan parasites of wild marine salmon (*Salmo salar* L.) from the west coast of Norway with comments on their interactions with farmed salmon. *Aquaculture*, **98:** 311–318.

BROSNUM, T.M. & O'SHEA, M. L. (1996) Long-Term Improvements in Water Quality Due to Sewage Abatement in the Lower Hudson River. *Estuaries*, **19:** (4) 890–900.

BROWNE, C., EDWARDS, S., ESCANDELL, M., FALCOUS, K. & JACKSON, T. (2001a) Rocky Shore Survey of Dunmanus Bay 1981–2000. Sherkin Island Marine Station (Unpublished report).

BROWNE, C., EDWARDS, S., ESCANDELL, M., FALCOUS, K. & JACKSON, T.

(2001b) Rocky Shore Survey of East of Baltimore 1981–2000. Sherkin Island Marine Station (Unpublished report).

BUCKEL, J. A., CONOVER, D.O., STEINBURG, N.D. & McKOWN, K.A. (1999) Impact of Age-0 Bluefish (*Pomatomus saltatrix*) Predation on Age-0 fishes in the Hudson River Estuary: Evidence for density-dependent loss of Juvenile Striped Bass (*Marone saxatilis*) *Canadian Journal of Fisheries Science.* **56:** 275–287.

BUCKEL, J.A. & McKOWN, K.A. (2002) Competition Between Juvenile Striped Bass and Bluefish: Resource Partitioning and Growth Rate. *Marine Ecology Progress Series* Vol **234:** 191–204.

BUCKLAND, S.T. & BREIWICK, J.M. (2002) Estimated trends in abundance of eastern Pacific gray whales from shore counts (1967/68 to 1995/96). *Journal of Cetacean Research and Management* **4:** (1) 41–8.

BUCKLAND, S.T., ANDERSON, D.R., BURNHAM, K.P., LAAKE, J.L., BORCHERS, D.L. & THOMAS, L. (2001) *Introduction to Distance Sampling: Estimating Abundance of Biological Populations.* Oxford University Press, Oxford, UK. vi+xv+432pp.

CAMPBELL, R.N. & WILLIAMSON, R.B. (1979) The fishes of the inland waters of the Outer Hebrides. In: *Proceedings of the Royal Society of Edinburgh* 'The Natural Environment of the Outer Hebrides'

CARSON, R. (1962) *Silent Spring.* The Riverside Press, Cambridge MA, USA. 368pp.

CASEY, S., MOORE, N., RYAN, L., MERNE, O.J., COVENEY, J.A. & DEL NEVO, A. (1995) The Roseate Tern conservation project on Rockabill, Co. Dublin: a six year review 1989–1994. *Irish Birds* **5:** 251–264.

CENTRAL HUDSON GAS & ELECTRIC, Consolidated Edison Company of NY, Inc., New York Power Authority, and Orange and Rockland Utilities, Inc. (1993) Draft Environmental Impact Statement for SPDES Permit for Bowline Point, Indian Point 2&3, and Roseton Electric Generating Stations. SPDES Permits: NY0008231; NY0004472; and NY0005711.

CHAMP, W.S.T. (1977) Trophic status of fishery lakes. In: *Lake Pollution Eutrophication,.* Downey, W.K. and Ni Uid, G. (eds), Dublin, Stationary Office. 65–78.

CHAMP, W.S.T. (1979) Eutrophication and brown trout fisheries. *Salmon & Trout Magazine.* **215:** 47–51.

CHAMP, W.S.T. (1993) Lough Sheelin – a 'success' story. In: *Water of Life*, C.Mollan (ed.), Dublin, Royal Dublin Society, 154–162.

CHAMP, W.S.T. (1998) Phosphorus/Chlorophyll relationships in selected Irish lakes: Ecological consequences and suggested criteria for ecosystem management. In: *Eutrophication in Irish waters,* Wilson, J.G. (ed.), Royal Irish Academy, Dublin, pp.91–105.

CHAMP, W.S.T. (1998) Phosphorus/Chlorophyll relationships in Irish lakes: ecological consequences and suggested criteria for ecosystem management. In: Wilson, J. (ed), *Eutrophication in Irish Waters,* 91–105. Dublin, Royal Irish Academy,

CHAMP, W.S.T. (2003) The effects of Pollution on Freshwater Fish Stocks. In: *Tackling pollution of Inland and Coastal waters.* Proceedings Nineteenth Annual environmental Conference. Sherkin Island Marine Station, Co. Cork.

CLARK, J. (1999) Bantry, Bay, Cork Harbour and East of Baltimore. Sherkin Island Marine Station (Unpublished report).

CLARK, J., GWENIN, S., NETZER, J., TOMLIN, H. & WADE, T. (1997) The Sherkin Island Marine Station 1997 Rocky Shore Survey of West Cork. Sherkin Island Marine Station (Unpublished report).

COBB, K. M., CHARLES, C. D., CHENG, H. & EDWARDS, R. L. (2003) El Niño/Southern Oscillation and tropical Pacific climate during the last millennium *Nature* **424**: 271–276.

COLEBROOK, J. M. (1960) Continuous plankton records: methods of analysis, 1950–59. *Bulletin of Marine Ecology.* **5**: 51–64.

COLLETE, B. B. & KLEIN-MACPHEE, G. (editors). (2002) *Bigelow and Schroeder's Fishes of the Gulf of Maine, 3rd edition.* Smithsonian Institution Press. 748 pp.

COOPER, J.C., CONTELINO, F. R., CROOM, J.M. & SHAPOT, R. (1988) Science, Law and Hudson River Power Plants. A Case Study in Environmental Assessment. Overview of the Hudson River Estuary. *American Fisheries Society*, Monograph **4**: 11–24.

CRAMP, S., BOURNE, W. R.P. & SAUNDERS, D. (1974) *The Seabirds of Britain and Ireland.* Collins, London.

CRAPP, G.B. (1973) The Distribution and Abundance of Animals and Plants on the Rocky Shores of Bantry Bay. *Irish Fisheries Investigations. Series B*, **9**: 3–35.

CRAWFORD, C.M., MITCHELL, I.M. & MACLEOD, C.K.A. (2001) Video assessment of environmental impacts of salmon farms. *ICES Journal of Marine Science,* **58**: 445–452.

DALE, A.L. & DALE, B. (2002) Application of ecologically-based statistical treatments for micropalaeontology. In: *Quaternary environmental micropalaeontology,* (S. Haslett, ed.). Edward Arnold Ltd., London, pp. 259–286.

DALE, B. (1976) Cyst formation, sedimentation, and preservation: factors affecting dinoflagellate assemblages in Recent sediments from Trondheimsfjord, Norway. *Review of Palaeobotany and Palynology,* **22**: 39–60

DALE, B. (1983) Dinoflagellate resting cysts: "benthic plankton". In: Fryxell, G.A.(ed.), *Survival strategies of the algae.* Cambridge University Press, New York, pp. 69–135.

DALE, B. & DALE, A. (2002) Environmental Applications of Dinoflagellate cysts and Acritarchs. In: *Quaternary Environmental micropalaeontology,* (S. Haslett, ed.) Edward Arnold Ltd., London, pp. 207–240.

DALE, B., DALE, A.L., & JANSEN, F.J.H. (2002) Dinoflagellate cysts as environmental indicators in surface sediments from the Congo deep-sea fan and adjacent regions. *Palaeogeography, Palaeoclimatology, Palaeoecology.*: **185**: 309–338.

DALE, B. AND NORDBERG, K. (1993) Possible environmental factors regulating prehistoric and historic "blooms" of the toxic dinoflagellate *Gymnodinium catenatum* in the Kattegat-Skagerrak region of Scandinavia. In: Smayda, T.J. & Shimizu, Y. (eds.). Toxic Phytoplankton Blooms in the Sea. *Proceedings of the*

Fifth International Conference on Toxic Marine Phytoplankton, Newport Rhode Island, USA, 28 October – 1 November, 1991. Elsevier Science Publishers, Amsterdam.

DEPARTMENT OF COMMUNICATIONS, the Marine & Natural Resources. (2000) Monitoring Protocol No. 3: Offshore finfish farms-sea lice monitoring and control.

DEPARTMENT OF FISHERIES AND OCEANS, (2003) State of the Eastern Scotian Shelf Ecosystem. DFO Ecosystem Status Report 2003/004.

DEPARTMENT OF FISHERIES AND OCEANS, (2003) Georges Bank Scallop. DFO Science Stock Status Report 2003/038.

DEPARTMENT OF THE ENVIRONMENT & Local Government. (1998) Local Government (Water Pollution) Act 1977, (Water Quality Standards for Phosphorus) regulations, 1998. Statutory Instrument 258 of 1998. Government Supplies Agency, Dublin.

DEPARTMENT OF THE MARINE. (1992) Report of the sea trout working group 1992. A report to the Minister of the Marine, 1992.

DEPARTMENT OF THE MARINE. (1993) Report of the sea trout working group 1993. A report to the Minister of the Marine, Oct/Nov 1993.

DEPARTMENT OF THE MARINE. (1995) Report of the sea trout working group 1994. A report to the Minister of the Marine, Feb/Mar 1995.

DODD, V.A. & CHAMP, W.S.T. (1983) Environmental problems associated with intensive animal production units, with reference to the catchment area of Lough Sheelin. In: *Promise and performance.* Irish Environmental Policies Analysed. Eds J. Blackwell and F. Convery, REPC, Dublin, 111–129.

EDWARDS, S., BROWNE, C., ESCANDELL, M., FALCOUS, K., JACKSON, T. (2001) Bantry Bay Rocky Shore Report 1995–2000. Sherkin Island Marine Station (Unpublished report).

EGERTON, M. (1998) Comparison of species distribution on selected Sherkin Island Annual Seashore Monitoring Sites from 1995–1998. Sherkin Island Marine Station (Unpublished report).

EGERTON, M., DALEY, L., PARTRIDGE, J., DUFFY, S., CLARK, J. & FARNELL, T. (1998) The Sherkin Island Marine Station 1998 Rocky Shore Survey of West Cork. Sherkin Island Marine Station (Unpublished report).

ENVIRONMENTAL PROTECTION AGENCY (2001) *An Assessment of the Trophic Status of Estuaries and Bays in Ireland.* Prepared for the Department of the Environment and Local Government. Environmental Protection Agency, Wexford.

ENVIRONMENTAL PROTECTION AGENCY, Marine Institute, Radiological Protection Institute of Ireland, Met Éireann, National Parks and Wildlife (2003) *National Environmental Monitoring Programme for Transitional, Coastal and Marine Waters. A Discussion Document.* Environmental Protection Agency, Wexford.

ETTER, M.L. (1996) Recent Changes in the Management and Handling of DFO Landings and Effort Data in the Scotia-Fundy Fisheries – Maritimes Region. In: *Canadian Technical Report of Fisheries and Aquatic Sciences.* **2100:** vii + 247 pp.

EUROPEAN COUNCIL (1976) Council Directive 67/464 of 4 May 1976 on pollution caused by certain substances discharged into the aquatic environment of the Community. *Official Journal of the European Communities* No L. 129/23.

EUROPEAN COUNCIL (1979) Directive 79/409/EEC of 2 April 1979 concerning the conservation of wild birds. *Official Journal of the European Communities*: L 103/25.4.1979 p.1.

EUROPEAN COUNCIL (1991a) Council Directive of 21 May 1991 concerning urban waste water treatment. 91/271/EEC *Official Journal of the European Communities*: L 135/30.5.1991 p.40.

EUROPEAN COUNCIL (1991b) Council Directive of 12 December 1991 concerning the protection of waters against pollution caused by nitrates from agricultural sources. 91/676/EEC *Official Journal of the European Communities*: L 375/31.12.1991 p.1.

EUROPEAN COUNCIL (1996) Directive 1996/61/EC of 24 September 1996 concerning integrated pollution prevention and control. *Official Journal of the European Communities*: No L 257, 10 October 1996, pp 26-40.

EUROPEAN COUNCIL (2000) Directive 2000/60/EC of the European Parliament and of the Council of 23 October 2000 establishing a framework for Community action in the field of water policy. *Official Journal of the European Communities,* No L 327, 22 December 2000, pp 1–72.

EVANS, P.G.H. & HAMMOND, P.S. 2004. Monitoring cetaceans in European waters. *Mammal Review,* **34:** 131–56.

FALCOUS, K., BROWNE, C., EDWARDS, S., ESCANDELL, M., JACKSON, T. (2001a) Rocky Shore Survey of Roaringwater Bay 1981–2000. Sherkin Island Marine Station (Unpublished report).

FALCOUS, K., BROWNE, C., EDWARDS, S., ESCANDELL, M., JACKSON, T. (2001b) 1995–2000 Rocky Shore Survey of Cork Harbour. Sherkin Island Marine Station (Unpublished report).

FANNING, P. & CASTONGUAY, M. (2003) The Comparative Dynamics of Exploited Ecosystems in the Northwest Atlantic Project. Symposium 16: Structure and Function of Continental Ecosystems : then and now, *133rd Annual Meeting of the American Fisheries Society,* Quebec City, Canada.

FINSTAD, B., BJORN, P. A., GRIMNES, A. & HVIDSTEN, N.A. (2000) Laboratory and field investigations of salmon lice [*Lepeophtheirus salmonis* (Krøyer)] infestation on Atlantic salmon (*Salmo salar* L.) post-smolts. *Aquaculture Research,* **31**: 795–803.

FITZMAURICE, P & GREEN, P. (2000) Results from Tagging of Blue Shark in Irish Waters. *The Irish Scientist Millennium Year Book.*

FOY, R.H., CHAMP, W.S.T. & GIBSON, C.E. (1996) The effectiveness of restricting phosphorus loadings from sewage treatment works as a means of controlling eutrophication in Irish lakes. In: Giller, P.S. and Myers, A.A. (eds), *Disturbance and Recovery in Ecological Systems,* 135–154. Dublin, Royal Irish Academy.

GARGAN, P.G., TULLY, O., & POOLE. R. (2003) The Relationship Between Sea Lice Infestation, Sea Lice Production And Sea Trout Survival In Ireland, 1992–2001. In: *"Salmon on the Edge"* (ed. D. Mills) Proceedings of The 6th International

Atlantic Salmon Symposium, Edinburgh, UK, July 2002, Chapter 10, pp. 119–135. Atlantic Salmon Trust/Atlantic Salmon Federation.

GARNER, G.W., ARMSTRUP, S.C., LAAKE, J.L., MANLY, B.F.J., MCDONALD, L.L. & ROBERTSON, D.G. (eds.). (1999) *Marine Mammal Survey and Assessment Methods.* Balkema, Rotterdam.

GERRODETTE, T. (1987) A power analysis for detecting trends. *Ecology* **68**: (5) 1,364–72.

GRANT, A. N. & TREASURER, J. (1993) The effects of fallowing on caligid infestations on farmed Atlantic salmon (*Salmo salar* L.) in Scotland. In: Boxshall, G. A. & Defaye, D. (eds), *Pathogens of wild and farmed fish: sea lice,* pp. 255–260. Ellis Horwood Ltd., West Sussex, United Kingdom.

HALL, M.A. & DONOVAN, G.P. (2002) Environmentalists, fishermen, cetaceans and fish: is there a balance and can science find it? pp. 491–521. In: P.G. Evans and J.A. Raga (eds.) *Marine mammals: biology and conservation.* Kluwer Academic/Plenum Publishers, New York.

HAMMOND, P.S. (1986) Estimating the size of naturally marked whale populations using capture-recapture techniques. *Report of the International Whaling Commission* (special issue) **8**: 253–82.

HAMMOND, P.S. & DONOVAN, G. (In press) Special Issue on the Revised Management Procedure. *Journal of Cetacean Research and Management* (special issue) **3**.

HAMMOND, P.S., MIZROCH, S.A. & DONOVAN, G.P. (eds.). (1990) *Report of the International Whaling Commission (Special Issue 12). Individual Recognition of Cetaceans: Use of Photo-Identification and Other Techniques to Estimate Population Parameters.* International Whaling Commission, Cambridge, UK. [vi]+440pp.

HAMMOND, P., BENKE, H., BERGGREN, P., BORCHERS, D.L., BUCKLAND, S.T., COLLET, A., HEIDE-JORGENSEN, M.P., HEIMLICH-BORAN, S., HIBY, A.R., LEOPOLD, M. & ØIEN, N. (2002) Abundance of harbour porpoises and other cetaceans in the North Sea and adjacent waters. *Journal of Applied Ecology.* **39**: 361–76.

HANNON, C., BERROW, S.D. & NEWTON, S. (1997) The status and distribution of breeding Sandwich *Sterna sandvicensis*, Roseate *S. dougallii*, Common *S. hirundo*, Arctic *S. paradisaea* and Little Terns *S. albifrons* in Ireland in 1995. *Irish Birds* **6**: 1–22.

HARDIN, G. (1968) The Tragedy of the Commons. *Science*, **162**: 1243.

HARDY, A.C. (1935) The continuous plankton recorder. A new method of survey. *Rapports et Proces-verbaux des Reunions. Conseil International pour l'Exploration de la Mer,* **95**: 35–77.

HARDY, A. (1956) *The Open Sea*; Fisher, J., Gilmour, J., Huxley, J., Davies, M. & Hosking, E: Eds, Collins, London.

HAWKEN, P. (1993) *The Ecology of Commerce.* Harper Business, New York. 250 pp.

HASLETT, S.K. (ed.) (2002) *Quaternary environmental micropalaeontology.* Edward Arnold Ltd., London.

HEIMBUCH, D. G., DUNNING, D. J. & YOUNG J. R. (1992) Estuarine Research in the 1980s. *The Hudson River Environmental Society Seventh Symposium on Hudson River Ecology.* State University of New York Press, Albany. 376–391.

HELLAWELL, J. M. (1978) Biological surveillance of rivers: A biological monitoring handbook. A collaborative production between Natural Environment Research Council, Water Research Centre and Regional Water Authorities in the UK. *Water Research Centre*, 332 pp.

HEUCH, P. A., NORDHAGEN, J. R. & SCHRAM, T. A. (2000) Egg production in the salmon louse [*Lepeophtheirus salmonis* (Krøyer)] in relation to origin and water temperature. *Aquaculture Research*, **31**: 805–814.

HIBY, A.R. & HAMMOND, P.S. (1989) Survey techniques for estimating abundance of cetaceans. *Report of the International Whaling Commission* (special issue) **11**: 47–80.

HOGANS, W. E. & TRUDEAU, D. J. (1989) Preliminary studies on the biology of sea lice, *Caligus elongatus, Caligus curtus* and *Lepeophtheirus salmonis* (Copepoda: Caligoida) parasitic on cage-cultured salmonids in the Lower Bay of Fundy. *Canadian Technical Report of Fisheries and Aquatic Sciences,* No. 1715: 14 pp.

HUNTER, J. (1997) The Rocky Shores of Sherkin Island 1987–1997. Sherkin Island Marine Station (Unpublished report).

HUTCHINSON, C.D. (1979) *Ireland's Wetlands and their Birds.* Irish Wildbird Conservancy, Dublin.

JACKSON, D. & COSTELLO, M. J. (1991) Dichlorvos and alternative sea lice treatments. *Aquaculture and the Environment*, Special publication No.16.

JACKSON, D. & MINCHIN, D. (1993) Lice infestation of farmed salmon in Ireland. In: *Pathogens of wild and farmed fish: sea lice.* Boxshall, G. and Defaye, D. Chichester, Ellis Horwood: 188–201.

JACKSON, D., HASETT, D. & COPLEY, L. (2002) Integrated lice management on Irish Salmon Farms. *Fish Veterinary Journal* **6**: 28–38.

JACKSON, T., BROWNE, C., EDWARDS, S., ESCANDELL, M. & FALCOUS, K. (2001) Sherkin Island Rocky Shore 1981–2000. Sherkin Island Marine Station (Unpublished report).

JOHANNESSEN, A. (1978) Early stages of *Lepeophtheirus salmonis* (Copepoda, Caligidae) *Sarsia* **63**: 169–176.

JONES, M. W., SOMMERVILLE, C. & BRON, J. (1990) The histopathology associated with the juvenile stages of *Lepeophtheirus salmonis* on the Atlantic salmon, *Salmo salar* L. *Journal of Fish Diseases*, **13**: 303–310.

JONSDOTTIR, H., BRON, J. E., WOOTTEN, R. & TURNBULL, J. F. (1992) The histopathology associated with the pre-adult and adult stages of *Lepeophtheirus salmonis* on the Atlantic salmon, *Salmo salar* L. *Journal of Fish Diseases,* **15**: 521–527.

KABATA, Z. (1974) Mouth and mode of feeding of Caligidae (Copepoda), parasites of fishes, as determined by light and scanning electron microscopy. *Journal of the Fisheries Research Board of Canada,* **31**: (10) 1583–1588.

KABATA, Z. (1979) *Parasitic Copepoda of British Fishes.* The Ray Society, London.

KAHNLE, A.& HATTALA, K. (1988) Bottom Trawl Survey of the Juvenile Fishes in the Hudson River, Summary Report for 1981–1986. *New York State Department of Environmental Conservation.* Mimeo.

KAHNLE, A., HATTALA, K. & STANG, D. (1988) Monitoring of the Commercial Gill Net Fishery for American Shad in the Hudson River Estuary. Summary Report for 1980–1986. *New York State Department of Environmental Conservation.* Mimeo.

KAHNLE, A., STANG, D., HATTALA, K. & MASON, W. (1988) Haul Seine Study of American Shad and Striped Bass Spawning Stocks in the Hudson River Estuary. Summary Report for 1982–1986. *New York State Department of Environmental Conservation.* Mimeo.

KENNEDY, M. & FITZMAURICE, P. (1971) The growth and food of brown trout *Salmo trutta* L. in Irish waters. *Proceedings of the Royal Irish Academy* **71B:** 269–352.

KENNEDY, P. G., RUTTLEDGE, R.F. & SCROOPE, C.F. (1954) *Birds of Ireland.* Oliver and Boyd, Edinburgh.

LATOUR, R. J., BRUSH, M. J. & BONZEK, C. F. (2003) Toward Ecosystem-Based Fisheries Management: Strategies for Multispecies Modeling and Associated Data Requirements. *Fisheries,* **10:** (9)10–22.

LEAHY, P. (1980) The effects of a dinoflagellate bloom on the invertebrate fauna of the sea-shore in Dunmanus Bay, Co. Cork, Ireland. *Journal of Sherkin Island* **1:** 119–125.

LLOYD, C.S. (1981) The seabirds of Great Saltee. *Irish Birds* **2:** 1–37.

LLOYD, C.S, TASKER, M.L. & PARTRIDGE, K. (1991) *The Status of Seabirds in Britain and Ireland.* Poyser, London.

LOMBORG, B. (2001) *The Skeptical Environmentalist: Measuring the Real State of the World.* Cambridge University Press, 515 pp.

MacKINNON, B. M. (1998) Host factors important in sea lice infections. ICES *Journal of Marine Science,* **55:** 188–192.

MANCRONI, W. J., DALEY, M. W. & DEY, W. (1989) Recent Trends and Patterns in Selected Water Quality Parameters in the Mid-Hudson River Estuary. Estuarine Research in the 1980s: *Hudson River Environmental Society, Seventh Symposium on Hudson River Ecology:* C. Lavett Smith, Editor. 59–81.

MARTIN, R. (ed) (2000) *Environmental micropalaeontology.* Kluwer Academic/Plenum Publishers, New York.

MASON, T.H. (1936) *The islands of Ireland.* Dublin.

MATSUOKA, K., ENSOR, P., HAKAMADA, T., SHIMADA, H., NISHIWAKI, S., KASAMATSU, F. & KATO, H. (2003) Overview of minke whale sightings surveys conducted on IWC/IDCR and SOWER Antarctic cruises from 1978/79 to 2000/01. *Journal of Cetacean Research and Management* **5:** (2) 173–201.

McGARRIGLE, M.L., CHAMP, W.S.T., NORTON, R., LARKIN, P. & MOORE, M. (1983) The trophic status of Lough Conn. An investigation into the causes of recent accelerated eutrophication. *Mayo County Council.* pp. 84.

McGARRIGLE, M., *et al.* (2002) *Water Quality in Ireland 1998–2000.* Environmental Protection Agency, Wexford.

McKOWN, K. A. (1991) Validation of the Hudson River Young-of-the-Year Striped Bass Indices. *Report of the Atlantic States Marine Fisheries Commission, Science and Statistical Committee.* Mimeo.

McKOWN, K. A. (2001) An Investigation of the 1999 Hudson River Striped Bass Spawning Success. A Study of Striped Bass in the Marine District of New York. *New York State Department of Environmental Conservation,* E. Setauket, NY. Completion Report for P.L. 89–304. Project No. AFC – 23.

McKOWN, K. A., & BRISCHLER, J.M. (2002) An Investigation of the Movements and Growth of the 2000 Hudson River Year Class. A Study of Striped Bass in the Marine District of New York. *New York State Department of Environmental Conservation,* E. Setauket, NY. Completion Report for P.L. 89–304. Project No. AFC – 24.

MERNE, O.J., BOERTMAN, D., BOYD, H., MITCHELL, C., Ó BRIAIN, M., REED, A. & SIGFUSSON, A. (1999) Light-bellied Brent Goose *Branta bernicla hrota*: Canada. In: Madsen, J., Cracknell, G. & Fox, A.D. (eds.) Goose populations of the Western Palearctic. A review of status and distribution. *Wetlands International Publ. No. 48*, Wetlands International, Wageningen, The Netherlands. National Environmental Research Institute, Ronde, Denmark. 298–311.

MERRIMAN, D. (1941) Studies on the Striped Bass (*Roccus saxatilis*) on the Atlantic Coast. *Fishery Bulletin* **50**: 1–77.

MIGUS, S. & MOORE, C. (2002) Rocky Shore Survey of Sherkin Island 1981–2000. Sherkin Island Marine Station (Unpublished report).

MITCHELL, P.I., NEWTON, S.F., RATCLIFFE, N. & DUNN, T.E. (2004) *Seabird Populations of Britain and Ireland.* Poyser, London.

MORISETTE, L., HAMMILL, M. & SAVENKOFF, C. (2003) Trophic roles of Marine Mammals in the Gulf of St. Lawrence Symposium 16: Structure and Function of Continental Ecosystems : then and now, *133rd Annual Meeting of the American Fisheries Society,* Quebec City, Canada.

NEILL, M. (2003). *River Water Quality in South-east Ireland, 2002.* A report commissioned by the County Councils of Carlow, Kilkenny, Laois, Tipperary (N&S), Waterford and Wexford and by Waterford City Council. Environmental Protection Agency, Kilkenny.

NELSON-SMITH, A. (1967) Marine Biology of Milford Haven: The distribution of littoral plants and animals. *Field Studies* **2**: (4) 435–477.

NOLAN, D. T., REILLY, P. & WENDELAAR BONGA, S. E. (1999) Infection with low numbers of the sea louse *Lepeophtheirus salmonis* (Krøyer) induces stress-related effects in post-smolt Atlantic salmon (*Salmo salar* L.) *Canadian Journal of Fisheries and Aquatic Science,* **56**: 947–959.

NOLAN, D. T., RUANE, N. M., VAN DER HEIJDEN, Y., QUABIUS, E. S., COSTELLOE, J. & WENDELAAR BONGA, S. E. (2000) Juvenile *Lepeophtheirus salmonis* (Krøyer) affect the skin and gills of rainbow trout *Oncorhynchus mykiss* (Walbaum) and the host response to a handling procedure. *Aquaculture Research*, **31**: 823–833.

NORDHAGEN, J. R. (1997) Livhistorie og morfologi til lakselus (*Lepeophtheirus salmonis*) fra villaks og oppdrettslaks. *Cand. Scient. Thesis, University of Oslo.*

NORDHAGEN, J. R., HEUCH, P. A. & SCHRAM, T. A. (2000) Size as indicator of origin of salmon lice *Lepeophtheirus salmonis* (Copepoda: Caligidae) *Contributions to Zoology*, **69:** (1/2) 99–108.

NEW YORK STATE DEPARTMENT OF ENVIRONMENTAL CONSERVATION. (2003) Final Environmental Impact Statement for SPDES Permits for the Roseton 1&2, Bowline 1&2 and Indian Point 2&3 Steam Electric Generating Stations, Orange, Rockland, and Westchester Counties. *New York State Department of Environmental Conservation.* Albany, NY.

O'GRADY, M.F. (1981) A study of brown trout (*Salmo trutta* L.) populations in selected Irish lakes. *Doctoral thesis,* National University of Ireland.

O'GRADY, M.F. (1983) A technique for estimating brown trout (*Salmo trutta* L.) populations in Irish Lakes. In: *Advances in Fish Biology in Ireland, Irish Fisheries Investigation, Series A*, No. **23:** 43–46.

O'GRADY, M.F., DELANTY, K. (2001) A review of changes in the fish stocks of Loughs' Conn and Cullin over time (1978–2001) and Recommendations in Relation to the Long-term management of these lakes and the river Moy as salmonid fisheries. *Central Fisheries Board.*

Organisation for Economic Co-operation and Development (OECD) (1982) *Eutrophication of waters. Monitoring, assessment and control.* Paris. OECD.

PALSBØLL, P.J., ALLEN, J., BÉRUBÉ, M., CLAPHAM, P.J., FEDDERSEN, T.P., HAMMOND, P.S., HUDSON, R.R., JÓRGENSEN, H., KATONA, S., LARSEN, A.H., LARSEN, F., LIEN, J., MATTILA, D.K., SIGURJÓNSSON, J., SEARS, R., SMITH, T., SPONER, R., STEVICK, P. & ØIEN, N. (1997) Genetic tagging of humpback whales. *Nature,* London. **388:** 767–9.

PARKER, M. (1980) Red tides. Fisheries Seminar Series, No 1. *Department of Fisheries,* Dublin.

PATRICK, D, ROBINSON, M, CROWE, O & NEWTON, S. (2002) Rockabill Tern Report 2002. *BirdWatch Ireland Conservation Report* No. 02/5.

PERRY, K.W. & WARBURTON, S.W. (1976) *The birds and flowers of the Saltee Islands.* Belfast.

PIASECKI, W. (1996) The developmental stages of *Caligus elongatus* von Nordmann, Copepoda: Caligidae. *Canadian Journal of Zoology,* **74:** 1459–1478.

PIASECKI, W. & MACKINNON, B. M. (1995) Life cycle of a sea louse, *Caligus elongatus* von Nordmann, 1832 (Crustacea, Copepoda, Siphonostomatoida) *Canadian Journal of Zoology,* **73:** 74–82.

PIKE, A. W. (1989) Sea lice – major pathogens of farmed Atlantic salmon. *Parasitology Today,* **5:** 291–297.

PLATT, T. C., FUENTES-YACO & FRANK, K.T. (2003) Spring algal bloom and larval fish survival. *Nature.* **423:** 398–399.

RAFTERY, A.E. & ZEH, J.E. (1998) Estimating bowhead whale population size and rate of increase from the 1993 census. *Journal of the American Statistical Association.* **93:** 451–63.

RAMSAR CONVENTION (1971) *Convention on Wetlands of International*

Importance: The Final Act of the International Conference on Conservation of Wetlands and Waterfowl, Ramsar, Iran, 1971.

RATCLIFF, J. (1999) Changes in the Community Structure of the Rocky Shore in Dunmanus Bay, as indicated by the changing abundance of fourteen selected species. Sherkin Island Marine Station (Unpublished report).

RICHARDS, R., RAGO, A. & RAGO, P. J. (1999) A case History of Effective Management: Chesapeake Bay Striped Bass. *North American Journal of Fishery Management* **19**: 156–175.

RITCHIE, G. (1993) Studies on the reproductive biology of Lepeophtheirus salmonis (Krøyer, 1837) on Atlantic salmon (*Salmo salar* L.) *PhD thesis,* Department of Zoology, University of Aberdeen.

ROBERT, G., BLACK, G.A.P., BUTLER M.A.E. & SMITH, S.J. (2000) Georges Bank Scallop Stock assessment – 1999. *DFO Canadian Stock Assessment Secretariat Research Document* 2000/016, 25 pp+fig.

ROCHE, R. & MERNE, O.J. (1977) *Saltees – Islands of birds and legends.* O'Brien Press, Dublin.

RODEN, C.M., RYAN, T. & LENNON, H.J. (1980) Observations on the 1978 red tide in Roaringwater Bay, Co. Cork. *Journal of Sherkin Island* **1**: 105–119.

RODEN, C.M., LENNON, H.J., MOONEY, E., LEAHY, P. & LART, W. (1981) Red tides, water stratification and phytoplankton species succession around Sherkin Island, south-west Ireland, in 1979. *Journal of Sherkin Island* **2**: 50–68.

ROMAN, J & PALUMBI, S. R. (2003) Whales Before Whaling in the North Atlantic. *Science*. 25 July, 2003; **301**: 508–510. (in Reports)

RUTTLEDGE, R.F. (1963) Migrant and other birds of Great Saltee, Co. Wexford. *Proceedings of the Royal Irish Academy.* Dublin. Section B, No. **4**: 71–86.

QUINN. T. & DERISO, R. (1999) *Quantitative Fish Dynamics (Biological Resource Management Series)* Oxford University Press. 560 pp.

SAINSBURY, K.J., PUNT, A.A. & SMITH, A.D.M. (2000) Design of operational management strategies for achieving fisheries ecosystem objectives. *International Council for the Exploration of the Sea (ICES) Journal of Marine Science* **57**: 731–741.

SAINSBURY, K.J. & SUMAILA, U. R. (2003) Incorporating Ecosystem Objectives Into Management of Sustainable Marine Fisheries, Including 'Best Practices' Reference Points and Use of Marine Protected Areas. In: Sinclair and Valdimarsson, ed., *Responsible Fisheries in the Marine Ecosystem,* FAO and CABI Publishing, 426 pp.

SCHRAM, T. A. (1993) Supplementary descriptions of the developmental stages of *Lepeophtheirus salmonis* (Krøyer, 1837) (Copepoda: Caligidae) In: Boxshall, G. A. & Defaye, D. (eds), *Pathogens of wild and farmed fish: sea lice,* pp. 30–47. Ellis Horwood Ltd., West Sussex, United Kingdom.

SHEPPARD, R. (1993) *Ireland's Wetland Wealth.* Irish Wildbird Conservancy, Dublin.

SLOAN, R. J., STANG, D. & O'CONNELL, E. A. (1988) PCB in Hudson River Striped Bass. Ten Years of Monitoring. *Technical Report 88–2 (BEP)* New York State Department of Environmental Conservation. 36 pp.

SLOAN, R. & HATTALA, K. A. (1991) Temporal and Spatial Aspects of PCB Contamination in Hudson River Striped Bass. *Technical Report 91–2 (BEP)* New York State Department of Environmental Conservation. 97 pp.

SLOAN, R.; YOUNG, B. & HATTALA, K. (1995) PCB Paradigms for Striped Bass in New York State. *Technical Report 95–1 (BEP)* New York State Department of Environmental Conservation. 117 pp.

SMAYDA, T. (1990) Novel and nuisance phytoplankton blooms in the sea: evidence for a global epidemic. In: *Toxic Marine Phytoplankton* (Graneli, E., Sundstrøm, B., Edler, L. & Anderson, D.M., eds). Elsevier, New York, pp. 29–40.

SMITH, T.D., ALLEN, J., CLAPHAM, P.J., HAMMOND, P.S., KATONA, S., LARSEN, F., LIEN, J., MATTILA, D., PALSBØLL, P.J., SIGURJÓNSSON, J., STEVICK, P.T. & ØIEN, N. (1999) An ocean-basin-wide mark-recapture study of the North Atlantic humpback whale (*Megaptera novaeangliae*). *Marine Mammal Science* **15:** (1) 1–32.

SQUIRES, D. F. (1983) *The Ocean Dumping Quandary: Waste Disposal in the New York Bight State.* Univ of New York Press, NY.

SUAREZ, M. (2003) Downriver migration of juvenile Hudson River Striped Bass: Patterns, Processes and Comparison of Habitat Quality. *Masters Thesis,* SUNY Stony Brook.

SUSKOWSKI, D. J., & WALDMAN, J.R. (1996) The Hudson River Estuary: Introduction to the Dedicated Issue. *Estuaries,* **19:** (4) 757–758.

SWEENEY, J., DONNELLY, A., MCELWAIN, L., AND JONES, M. (2002) *Climate Change Indicators for Ireland.* ERTDI Report Series No. 2. Environmental Protection Agency, Wexford.

SWEENEY, J. *et al.* (2003) *Climate Change – Scenarios & Impacts for Ireland.* ERTDI Report Series No. 15. Environmental Protection Agency, Wexford.

SY, A., RHEIN, M., LAZIER, J.R.N., KOLTERMANN, K.P., MEINCKE, J., PUTZKA, A. & BERSCH, M. (1997) Surprisingly rapid spreading of newly formed intermediate waters across the North Atlantic Ocean. *Nature* **386**: 675–679.

TAYLOR, S. & HUNTER, J. (1996) A survey of the rocky shores of West Cork 1995–1996. Sherkin Island Marine Station (Unpublished report).

THOMPSON, W. (1851) *The Natural History of Ireland.* Vol **3**. Reeve & Benham, London.

USSHER, R.J. & WARREN, R. 1900. *Birds of Ireland.* Gurney & Jackson, London.

VERSAR INC. (1990) Atlantic States Marine Fisheries Commission, Fishery Report No. 16 of *Atlantic States Marine Fisheries Commission. Source Document for the Supplement to Striped Bass FMP Amendment No. 4.* 363 pp.

VOLLENWEIDER, R.A. (1971) *Scientific fundamentals of the eutrophication of lakes and flowing waters, with particular reference to nitrogen and phosphorus as factors in eutrophication.* Paris. OECD.

WADSWORTH, S., GRANT, A. & TREASURER, J. (1998) A strategic approach to lice control. *Fish Farmer,* March/April: 8–9.

WALSH, J.J. & STEIDINGER, K.A. (2001) Sahara dust and Florida red tides: The cyanophyte connection. *Journal of Geophysical Research* **106:** (C6) 11597–11612.

WHILDE, A. (1985) The 1984 All-Ireland Tern Survey. *Irish Birds* **3:** 1–32.

WOOTEN, R., SMITH, J. W. & NEEDHAM, E. A. (1982) Aspects of the biology of the parasitic copepods *Lepeophtheirus salmonis* and *Caligus elongatus* on farmed salmonids, and their treatment. *Proceedings of the Royal Society of Edinburgh,* **81B:** 185–197.

YOUNG, B. H. (1976) A Study of Striped Bass in the Marine District of New York. *New York State Department of Environment Conservation,* E. Setauket, NY. Completion Report for P.L. 89–304. Project AFC – 8. Mimeo. 64 pp.

YOUNG, B. H. (1980) A Study of Striped Bass in the Marine District of New York. *New York State Department of Environment Conservation,* E. Setauket, NY. Completion Report for P.L. 89–304. Project AFC – 9. Mimeo. 89 pp.

YOUNG, B. H. (1986) A Study of Striped Bass in the Marine District of New York. *New York State Department of Environment Conservation,* E. Setauket, NY. Completion Report for P.L. 89–304. AFC – 12. Mimeo. 73 pp.

INDEX

a) Subjects

Note that entries relating to Ireland are grouped under 'Ireland'

A

acid deposition / acid rain 25, 26, 38, **46,** 50
Afghanistan 53
Africa, West 83, 108
agriculture 74, 180, 182
    livestock 187
air
    quality 15, **38**
    temperature 192
algae, see also under their generic or common names 46, 69, 75, 117, 118, 130, 183
algal blooms – see also under HABS and phytoplankton 66, **67, 107,** 181, 183
algal sporelings 126
amateurs, monitoring by **11,** 49, **195,** 197
Amazon rainforest 30
American Fisheries Society 135, 139
analytical power (statistics) 149
anglers 47, 49, 74
    catches, see also census of creels 67
Anglesey 41
animal behaviour 184
animal waste management 187
Antarctica 26, 38, 47, 169
Anti-fouling paint 47
aquaculture 160, 184
aquatic resources 193, 195
archives
    natural 108
Arctic 82, 162
assessment
    biological 37
assimilation capacity, of ecosystems 49, 197
Atlantic Ocean 49, 75, 78, 88, 111, 132, 157, 190
Atlantic States Marine Fisheries Commission 132, 136
atmosphere 26, 152
    changes in 11, 47
    chemistry 42
    emissions to 38, 108
avalanche 61
Azores, Islands 75

B

bacteria 46, 187
Baja California, Mexico 162
Balkans 53
Baltimore, Maryland, USA 135
bathing waters 46
Bay of Biscay 86
Bay of Fundy 190
Beaufort Dyke 53
Bedford Institute of Oceanography, Canada 151
Benguela Current 109
Benguela Front 110
benthic 141
    communities 154
    environment 152
Bering-Chukchi-Beaufort Seas 169
biochemical oxygen demand (BOD) 135
biochemistry 40
biodiversity 49, 61, 140
biogeography 109
bioluminescence 157
biomagnification 47
biomarkers 112
birds (see also under named species) 10, 47, 184, 188
    census 9, 79, **81,** 196
    distribution 79
    migration 11, 79
    observatories 79
    ringing 81, 85
    satellite tracking 81
    status 79
birth defects 53
bogflows 61, 64
Booker McConnell 87
Boston 190
botany 40
botulism, avian 85
British Antarctic Survey **42,** 52
    Halley Station 43
British Trust for Ornithology 9, 12
Broadbalk experiment **41**

C

caesium 64
Calicide 95
California 64
Canada 158, 159
    Atlantic 149, 150
    High Arctic 81
Canary Islands 57, 75
cancer 1, 53
Cape Ann, USA 159
carbon dioxide 8, 47, 52, 152
carbon monoxide 38
Caribbean 56
Carnan, Loch 89
carrying capacity, of the Earth 152
catch and release (fishery) 72
catch per unit effort 70, 71, 137
catch quotas 200
census
    birds 85, 196
    creels 68
    land-based, of whales 169
    mark-recapture 170
    migrating shoreline birds 196
Cetaceans **161**
    abundance 163
    catches 164
        direct 163
        incidental 162, 163

habitat
    loss of *162*
    hunting of *162*
    management objectives *163*
    prey species
        overfishing *162*
    reproductive stressors *163*
    ship-strikes *162*
changes, tracking *187*
chemical analysis *15*
chemicals *55*
    industrial *134*
    xenobiotic *49*
chemical weapons, dumping of *53*
chemistry of sediments *194*
chemotherapeutants *95*
Chernobyl *27*, **61**
cherry blossom *12*
Chesapeake Bay *132*
children, abnormalities in *53*
chlorofluorocarbons *42, 52*
chlorophyll, chlorophyll-A *67, 69, 71, 72, 140, 152*
cherry blossom, ancient records of *12*
clam fishers ('clammers') *158*
clarity of water, see also transparency *26, 71*
Clean Water Act, USA *199*
climate *26*, **62**, *110*
    change *11*, **38**, *39, 47, 48, 49, 62, 64, 153, 155, 175*, **183**, *190*
        indicators of *183*
    monitoring of **8**
climatology *40, 52*
coastal
    defences *19*
    waters
        releases into *187*
    zones *187*
coccoliths, see micro-fossils
communication **31**, *64*
communities *152*
    biotic *177*
    structure *194, 198*
Comparative Dynamics of Exploited Ecosystems in the Northwest Atlantic Project *154*
conservation *80*
    Cetaceans **161**
    marine ecosystems *159, 184*
continental drift *9*
continental shelf *141, 153*
Continuous Plankton Recorder *7, 49, 140*
Copenhagen *27*
copper *27*
coral reefs *189*
crops (vegetable) *61*
    genetically modified *54*
currents *152*
cycles of species abundance *130*
cypermethrin *95*

D

data **21, 22**
    accessibility **26**, *138, 164, 176*
    accuracy *15*, **23**, *198*
    analysis *164,*
    bank *12*, **22**
    base **22**, *77, 148, 158*

    collection practices *188, 193, 195*
    compatibility *65*
    environmental
        for decision-making **189**
    interpretation **22, 28,** *54*
    map-based *117*
    outlier *28, 29*
    photographic *117, 199*
    precision *15*, **23**
    quality control *150, 168*
    quality required *23*, **198**
    relevance *15*
    reliability *15*
    requirements *164*
    set *11, 28, 78, 111, 133, 135, 148, 149, 183, 188, 193, 196*
    time-series of *139, 141, 159*
    validation *19*
    video-recorded *152*
decision-making *199, 200*
dendrochronology **27**
Department of Health, UK *51*
depleted uranium *53*
diarrheic shellfish poisoning (DSP) *107, 143*
diatoms, see also micro-fossils *46, 108, 126, 140*
    frustules *42*
dinoflagellate cysts, see also micro-fossils *108*
dinophysistoxin-1 (DTX1) *143*
dinosaurs, extinction of *27*
Disease *183*
    Creutzfeldt-Jacob *52*
    hereditary *52*
    Mad Cow *52, 55*
    waterborne *54*
DNA (deoxyribo-nucleic acid), in whales *186*
drainage systems *194*
dredging
    materials, management of *194*

E

earthquakes *9, 56*
Eastern Scotian Shelf, Canada *154*
eco-epidemiology *48*
ecological classification *74*
    and status *72, 77*
ecosystems *44, 151*
    aquatic *38, 87*
    destruction of *47*
    evolution of *153*
    management of *9*
    marine, conservation of *159, 185*
    network of sites for monitoring *184*
    over-exploitation *47*
    replacement of *47*
education *49, 50*
El Niño *64, 108, 111, 186*
emamectin benzoate *95*
England
    Lake District *41*
environmental impact assessments *9, 31, 90, 136*, **200**
environmental impact statements *9, 200*
Environment for Europe Conference *37*
erosion of coastlines *62, 183*
estuaries, quality of *184, 200*
European Commission *7*

European Directives
    Birds *80, 82, 83*
    Dangerous Substances *72*
    Integrated Pollution Prevention and Control (IPPC) *178*
    Nitrates from Agricultural Sources *180*
    Urban Waste Water Treatment *180*
    Water Framework Directive *38, 48, 59, 72, 74, 77, 175, 176, 177, 178, 188*
European Environment Agency *32, 37*
European Marine Strategy *184*
European Sixth Environmental Action Programme *184*
Eutrophication *1, 47,* **67,** *68, 69, 75,* **110,** *112, 181*
    indicators **71**
    in fresh waters *68*
    in tidal waters *37, 180*
Excis *95*

F

Falklands
    War *43*
Faroes, Islands *75*
Federal Clean Water Act, USA **188**
Ferry, English Channel *55*
fertilizers *61, 187*
filter feeders *153*
Finland *49, 61*
finnock *73*
fish *47,* **66**
    as environmental indicators **74**
    catch-and-release *72*
    catches
        historic *186*
        rod *72*
        traps *196*
    juvenile, index of survival *197*
    movement orders *96*
    otoliths *27*
    scales *27*
    spawning *68, 190, 192, 195, 197*
    stock assessment *67, 77, 155, 197*
    tagging programmes *74*
    young of the year, abundance index *133, 136, 200*
Fish and Wildlife Service (USA) *136, 196*
fisheries management *46, 47, 77, 139, 189, 193*
fishery *48, 110,* **132,** *152, 154, 184, 186*
    coastal *134*
    closure of *135, 159, 200*
    collapse of *110, 154, 193*
    economic value *92*
    efficiency *153*
    groundfish *151, 154*
    habitat *194*
    invertebrate *152*
    landing records *153*
    lobster *154*
    pelagic *151*
    wild harvest *154, 191*
        economic value *49, 150*
Fishery Management Plan, striped bass *133*
fish farming **87**
    conditional fish movement orders *96*
    fallowing *95, 104*
    husbandry *95, 96*
    inspection, sampling *95, 99*
    mixed-generation sites *105*

Single Bay Management *96*
single-generation sites *105*
treatment of infestations
    bath *99*
    in-feed *99*
    synchronous *105*
    triggers for *96, 99*
fish processing *150*
fish stocks, and decline of *107*
flooding and flash-flooding *1, 57, 62, 194*
flood plain *19*
    protection of *19*
    sediments *57*
flora and fauna
    intertidal *41, 114*
    monitoring of *38*
Florida *108, 111*
fluorescence *157*
food
    chain *47, 61, 87, 157*
    quality of *61, 64*
    -web *47, 152, 159*
fossils, see also micro-fossils *26, 27, 108, 109, 185*
forests *30*
    tropical *16, 30*
    urban *198*
fossil fuel *47*
France *110*
Frederiksborg *27*
Freshwater Biological Association *41, 49*
    Windermere Laboratory *41*
Frierfjord *110*

G

Gaia Hypothesis **53**
genetic fingerprinting *170*
Geographic Information Systems (GIS) *50, 56, 198*
geology *40, 46, 50,* **56**
geomagnetism *40*
geomorphology *88*
George's Bank *153*
gill nets *67, 68, 70, 77, 136*
global
    change *107, 190*
    data sets *38*
    marine phytoplankton monitoring **112,** *113*
    warming **1, 2,** *42, 47, 51, 188*
greenhouse gases *8, 39, 42, 52, 63*
Greenland *81*
groundwater *59, 60, 64, 177*
    levels *60*
    protection schemes *59*
    quality *60*
Gulf of Mexico *111*
Gulf of St. Lawrence, Canada *154*
Gulf Stream *16*
Gulf War Syndrome *53*

H

habitats *177, 197*
    destruction of **1**
    loss of *194*
    management of *83*
    protecting the integrity of *177*

recovery of *10*
HABS (harmful algal blooms) *107, 111, 157, 158*
   aerosol-producing *108, 198*
   predictive indices *159*
'Harbor Watch', USA NGO *195*
harvest, sustainable *165*
hazardous waste sites, clean-up *199*
health, see public health
Hebrides, Outer *87*
historical baseline (flood-plain sediment) *57*
HMS Beagle *18*
HMS Challenger *18*
hormone replacement therapy *47*
hormones *47*
   sex, mimicry of **47**
   stress *93*
Hudson River Estuary Management Program (USA) **134,** *138*
hurricane *56*
hydrates **63**
hydrogeology *64*
hydrology / hydrologists *46*
hydrophones, use in monitoring whales *171*
hyperbaric chamber *159*
hypertrophic lakes *75*
hypoxia *135, 196*

I

Ice Age, see also Little Ice Age *57, 69*
ice cores *26*
   air trapped in Polar ice *42*
Iceland *81, 167*
ice *42, 47*
   movements *43*
   sheets *57, 167*
ichthyoplankton *137*
identification
   individual whales *170*
   trees *198*
industrial development *11*
industry *184*
   biotech *54*
insurance losses *56*
interested parties, see stakeholders
International Climate Panel *2*
International Whaling Commission *163*
   Antarctic cruises *172*
Iraq *53*

IRELAND *53, 75, 104, 175*
   Baltimore *115, 117*
   Bays
      Bantry *114, 116*
      Clew *62, 63, 98*
      Donegal *99*
      Dunmanus *116*
      Kenmare *99*
      Kilkieran *99*
      Long Island *114*
      Roaringwater *114, 116, 141*
   BirdWatch Ireland *9, 12, 83*
   Birr *39*
   Brent Goose Research Group *81*
   Cape Clear Island *79*
   Castlemaine Harbour *81*
   Connemara *72, 98*

Copeland Islands *79*
Cork City *57*
Cork Harbour *115, 117*
Countryside Bird Survey *80*
County
   Clare *82*
   Cork *79, 97*
   Derry *81*
   Donegal *79, 81, 82, 97*
   Down *79, 81, 82*
   Dublin *61, 82*
   Galway *82, 97*
   Kerry *81, 82, 85, 97*
   Louth *82*
   Mayo *61, 73, 82, 97*
   Sligo *82*
   Waterford *82*
   Wexford *39, 79, 82, 84*
   Wicklow *82*
Department of Communications, Marine and Natural Resources *96, 97, 177*
Department of the Environment, Heritage and Local Government *177*
Dublin City *38, 59*
Dublin Port Tunnel *59*
Environmental Protection Agency *52, 58, 59*
   Millennium Report *37*
   Office of Environmental Enforcement *178*
   Water Quality Report *37*
Environmental Protection Agency Act, 1992 175, *175, 176*
Erriff catchment *73*
Estuary
   Bandon *180*
   Barrow *180*
   Blackwater *181*
   Castletown *181*
   Broadmeadow *180*
   Feale *180, 181*
   Lee/Lough Mahon *181*
   Lee Upper (Tralee) *181*
   Liffey *180*
   Owennacurra *180*
   Slaney *180*
   Suir *180*
Fastnet Rock *49*
Fisheries Boards, Central and Regional *66, 70, 77, 97, 177*
Fisheries Conservancy Boards *66*
Gascanane Sound *141*
Geological Survey of *57, 177*
Great Saltee Island *79,* **84**
Inland Fisheries Trust *66, 74*
Irish Salmon Growers' Association *97*
Irish Wildbird Conservancy *80*
John F Kennedy Arboretum *39*
Keeragh Islands *85*
Killybegs Harbour *181*
Kilmore Quay *84*
Little Saltee Island *84*
Little Skellig Island *85*
Lough
   Arrow *66*
   Carlingford *83*
   Carra *66*
   Conn/Cullin *66, 70, 74*

Corrib *66, 68, 74*
Derravaragh *66*
Ennell *66, 67, 68, 72, 74, 75*
Foyle *81*
Inchiquin *66*
Leane *41*
Mask *66, 68*
Owel *66*
Sheelin *66, 68, 70, 72, 75*
Strangford *81, 82*
Malin Head *79*
Marine Institute *97, 176*
Marine Monitoring Forum *178*
Marine Sport Fish Tagging Programme 74
Maynooth *39*
Mayo, bogflow *61, 62*
Met Éireann *38, 176*
Monitoring programmes, National *99, 175*
    Air Quality (2000) *176*
    Estuarine and Coastal Water Quality (1996) *175, 176*
    Groundwater Quality (1997) *176*
    hydrometric *176*
    Lake Water Quality (2001) *176*
    River Water Quality (2002) *176*
    Sea Lice-Monitoring and Control Programme **95**
    tidal waters *175*
    Transitional, Coastal and Marine Waters (2003) **176**
Moycullen, Co. Galway *58*
National Botanic Garden *39*
National Climate Change Strategy *39*
National Parks & Wildlife Service *80, 82, 176*
National Seabed Survey *62*
National Waterfowl Census *79*
Office of Public Works *59*
Pollatomish, Co. Mayo *62*
Radiological Protection Institute *176*
Red List of Birds of Conservation Concern in Ireland *83, 86*
Refuge for Fauna *83*
River *175*
    Bush *49*
    Corrib *49*
    Shannon *181*
River Basin Management Plan *178*
Rockabill Lighthouse 82
Save Our Sea Trout *97*
Sea Trout Working Group *105*
Sherkin Island *116*
    Drolain Point *117*
    Globe Rocks *117, 119*
    Horseshoe Harbour *117, 118*
    Kinish Harbour *119*
    Kinish Narrows *117*
    Poulacurra *117, 119*
    Reenahoe *117, 119*
Sherkin Island Marine Station *36, 49, 55, 80, 107, 112,* **114, 140**
Silvermines, Co. Tipperary *58*
Tawnyard sub-catchment *73*
Tralee Bay *81*
Trinity College *39*
Tory Island *79*
Valentia Observatory *39*
Waste Management Act *178*
Western Gamefishing Association *97*

Westport Harbour *62*
Wexford Harbour *83*
Wetland Bird Survey *80, 81*
Whiddy Island *114*

Irish Sea *86*
Isle of Man *53*

J

Joint Assessment and Monitoring Programme (of OSPAR) *177*

K

kelt *73*
key species *135*
krill *169*
Kyoto Agreement *39, 64*

L

lakes *42*
    alkaline *66, 75*
    eutrophic *68*
    hypertrophic *75*
    mesotrophic *67, 68, 72, 75*
    monitoring of, including water levels *71*
landfills *58*
landslides *57, 58, 61, 62, 64*
    undersea *63*
leachate, from landfills *60*
lead *38, 53, 58*
lifecycle **93**
light *199*
limnology *106*
Little Ice Age *110*
local authorities *177*
Long Island Sound Study (USA) *138, 196*

M

machair *88*
magnetism *40*
Maine *158*
management *138, 152,* **164,** *184*
    objectives *164, 165*
Marine Harvest *90*
marine resources *184*
mark-recapture *170*
Massachusetts *158*
media **51,** *158*
Mediaeval warm period *2, 110*
metadata *26*
meteorological records *46, 49*
methane *58, 63*
micro-fossils *108, 109*
    coccoliths *108*
    diatoms *108*
    dinoflagellate cysts *108*
micropalaeontology **108**
Middle Ages *11*
migratory species *162, 197*
millinery trade *82*
mines and mining *58, 59*
    coal *58, 59*
Ministry of Defence, UK *53*

models, mathematical and modelling populations **42**, *138*, *139*
    as an aid to decision-making **192**
    calibration of *25*
    computer simulation *173*
Management Strategy Evaluation *152*
monitoring **45**, *99*, *159*
    activities licensed by authorities *178*
    air quality *38*
    bathing water *7*
    biological *32*, *74*
    birds **79**
    by amateurs *173*
    climate **8**
    compliance *11*, **29**, *46*
    drinking water *7*
    estuarine, coastal and marine water quality, including transitional waters *177*
    flora and fauna *38*
    groundwater *60*, *64*
    investigative programmes *178*
    networks *8*, *64*
    operational programmes *164*, *178*
    physical conditions *74*, *182*
    phytoplankton *159*
        global marine programme **112**
    programme, design of *163*
        spatial and temporal scale *177*
    rivers *48*, **152**, **153**
    schools, involvement *32*, *195*, *197*
    sea-lice **95**
    short-term **40**, *43*
    surveillance programmes *177*
    terminology *176*
    tidal waters *175*, *180*
    tiered *194*
    whales *162*
Mori *51*
Mount St Helens *27*

N

Narragansett Bay *42*
National Marine Fisheries Service (USA) *136*
National Pollution Discharge Elimination System (USA) *199*
Natural Environment Research Council (UK) *42*
natural resources, depletion of *37*
navigation *40*
neurotoxicity **158**
New England *157*
Newfoundland *154*
New York City *191*
New York/New Jersey Harbor Estuary Program (USA) *136*, *138*, *189*
nitrogen, oxides of *47*, *180*
Non-Governmental Organisations *196*
North Atlantic Oscillation *19*, *41*
North Sea *140*, *166*
Norway *110*
Nova Scotia, Province *153*
nuclear power *55*
nutrient *19*, *26*, *42*
    loading *69*, *135*, *181*
        estuaries *184*, *191*
    requirements *41*

O

oceans *11*
    changes in *11*
        ecosystem resources *185*
        warming *63*
Ocean Drilling Program *112*
oceanography *40*, *106*, *111*, *167*
oil, Kuwait crude *114*
oil rigs *169*
oil spills *44*, *86*
oil tanker *114*
okadaic acid *143*
Operation Seafarer *80*, *85*, *86*
oral tradition *149*
Organisation for Economic Co-operation & Development (OECD)
    Environmental Performance Review of Ireland *37*
organic matter *26*, *152*, *184*
organo-chlorine compounds *18*
ornithology *12*
Oslofjord *110*
Oslo & Paris Convention (OSPAR) *177*
otoliths *27*
otter **87**
    diet *89*
    holts *88*
    skins *88*
    spraints *88*
    tracking, using radiotelemetry *90*
    trapping *88*
overfishing *30*, *151*, *193*
over-grazing *61*
oxygen, dissolved *46*, *135*, *190*, *191*, *194*, *199*
ozone layer *47*, *152*, *195*
    depletion of / holes in **1**, **42**, **43**, *47*, *52*

P

Pacific coast *132*, *157*, *162*, *197*
paleontology
    micro- *108*
paralysis, respiratory *159*
Paris *59*
Parker River National Wildlife Refuge, MA, USA *158*
peach blossom, ancient records *12*
peat, peat bogs *26*, *42*, *88*
Peconic Estuary Program (USA) *138*
pesticides *10*, *47*, *88*, *187*
pests *183*
pH *25*, *46*, *50*, *88*
phenology **12**, *45*, **49**, *184*
phosphorescence, see bioluminescence
phosphorus *53*, *67*, *72*
    molybdate-reactive phosphate *69*
    removal of *181*
    total *67*, *69*, *71*, **75**
photosynthesis *8*
phytoplankton, see also plankton **106**, **112**, **140**, *157*
    dynamics of *147*
    seasonal variation in *190*
plankton *7*, *26*, *41*, *42*, *49*, *93*
    blooms, see algal blooms
    effects of atmospheric circulation on *41*
    long-term records of *49*
plant growth *41*

policy
  environmental *37*
pollen *26, 42*
pollutant loading *134*
pollution *1, 46, 48, 53, 64, 107, 110, 158, 162*
  control *187, 191*
  non-point sources *187*
  pathogenic *187*
  point-sources *15, 59, 187*
  water *48*, 187
polychlorinated biphenyls *135*
population *120*
  collapse *165*
  dynamics *145, 147*
  status assessment *163*
  trends (birds) *80, 85*
Portugal *110*
post-smolts *73*
potassium spectrometry *61*
power plants / power stations *47, 135, 160*
precautionary principle *54*
predators / predation *90, 154*
predator/prey relationships **126**
prediction **30**
productivity (birds) **81**
productivity, primary *152*
public health *38, 158, 159, 177, 187*
public pressure *199*

Q

quadrat *117*
Quebec *138, 158*

R

radar, satellite-based *59, 64*
radiation
  solar 42
  ultra-violet *1,* **42**
radiotelemetry, for tracking otters *90*
radon *58*
rainfall *41, 48, 50,* **194**
  management of *194*
  patterns of *38, 183*
  summer *183*
  winter *183*
Ramsar Convention *80*
real-estate values *187*
recommendations
  from the Workshop **32**
  other *184*
redds
  counts *67, 68*
red tides *141, 157, 190*
reference
  collections *16,* **17**
  conditions *72*
  sites **17**
  specimens *18*
refuge for fauna *83*
reproductive failures and successes *192*
resources
  conflicts over **186**
  management tools and decisions *200*
  recovery of *160*

renewable *193*
Rhode Island, USA *189*
ringing, birds *81, 85*
risk *33,* **51,** *200*
  assessment **45**
River (see also Ireland, River…)
  Congo *109*
  Delaware *132*
  Hudson **132,** *196*
  Roanoke *132*
  Saint John (Florida) *132*
  Saint John's (Canada) *132*
river **181**
  basin *38, 48, 178*
  flows *76*
  water temperatures *182*
'River Keeper', USA NGO *196*
rocky shores *16, 45, 49,* **114,** *130*
rod catches *72*
Rothamsted Research Institute *41*
  Broadbalk experiment *41*
Royal Meteorological Society *12*
Royal Society for the Protection of Birds *9, 12, 83*
run-off 30

S

salinity **134,** *152*
salmon farming *73, 87, 152, 183*
salt marshes *183, 194*
  plants *196*
samples *195*
  archives of *16, 141*
  biological **16**
  labelling of *16*
  preserving *141*
  reference collections of *198*
  storage of *17, 147*
sampling *67,* **95,** *189, 190, 195, 196*
  damage caused by *19, 20*
  design **23,** *48, 184*
  duration of **19,** *48, 192*
  'fit for purpose' *48*
  frequency **19,** *20, 23,* **24,** *48, 96*
  limitations *20*
  location **15**
  methods **20,** *48*
    standardised *25, 112, 117*
  Nansen bottle, use of *141*
  phytoplankton net, use of *141*
  quadrat, use of *117*
  reference sites *17*
  sites **15,** *48*
  stratified *80*
  transects, use of *117, 166*
  trial periods *20*
  variability *28*
sanitation *54*
satellite, monitoring by *42, 81, 153*
Scandinavia *46, 107, 110*
scanning, hyperspectral *61*
Scotia Fundy Fisheries Management Area (Canada) *150*
Scotland *46, 53, 87, 104*
sea angling *74*
  value of *74*
sea-bed charts *62*

Seabird 2000 *80, 86*
Seabird Colony Register *80*
Seafloor monitoring network *64*
sea-levels, rising *1, 10, 38, 47, 62, 64, 183*
sea lice **74, 92**
    control of *95, 99*
    infestations by **73,** *74*
    life cycles *93*
    monitoring of *95*
    treatment of, *96, 99, 105*
        by chemotherapeutants *99*
        by Single Bay Management Strategy *96*
sea mammals *154*
sea trout collapse *73, 77*
sea water *15*
    and density of *47*
Secchi disc *67, 199*
sedimentary record (of phytoplankton) **106**
sediments *26, 42, 108*
    chemicals in *42*
    coastal *62*
    lake *26, 27, 46*
sentinel species *187*
sewage discharge *134, 135, 190*
    abatement, see also wastewater treatment *135*
    outfall pipes *190*
sheep, grazing *61*
Sheilavaig, Loch *87, 89*
shellfish
    farms *187, 190*
        closure of *143, 188*
    industry *143*
    poisoning / toxicity *147,* **157, 158**
    spawning *190*
ships and shipping
    ballast water *107*
    coastguard *169*
    Erika 86
    ferries 169
    Prestige 86
    research 169
    Titanic 64
    Universe Leader, oil tanker *114*
    Viking *110*
    whalewatching *169*
'Silent Spring' **47**
size limits *200*
Slice *95*
smog *38*
smoke *38*
smoking *52*
smolts *92*
soil
    erosion *57, 64*
    permeability, sensitivity of trees to *61*
solar radiation *42*
sound
    velocity of, in ocean *63*
Southeast Area Marine Assessment Program (USA) *138*
Southern Ocean *41*
Southern Oscillation (El Niño) *186*
South Uist, Island *87*
Spain *110*
Special Protection Area (for birds) *82*
species
    abundance **127,** *129, 130, 137, 172, 189*

    changes *154, 183*
    density *118, 140, 152*
    diversity *130*
    key *135*
    migratory *188*
    productivity *130*
    richness **118**
    seasonal variation **119,** *189*
    sentinel *187*
St Lawrence, Canada *158*
stakeholders *14, 25, 164, 166*
standards *50, 72, 75*
    compliance *46*
statistics and statistical techniques *20,* **23,** *28, 41, 112, 192, 196*
    canonical correspondence analysis *109*
    Mann-Whitney U Test *99*
    power analysis *173*
streams, spawning *67*
stress hormones *93*
storms *1, 85, 190*
stormwater
    drains *194*
subsidence *59, 64*
sulphur dioxide *38, 47*
surveillance *5,* **45,** *186*
surveying *139*
    aerial *59, 64, 166*
        electromagnetic *60, 61*
        laser *64*
    epibenthic sled *137*
    geochemical *58*
    guidelines *168*
    hyperspectral *61*
    ichthyoplankton *137*
    remote-sensing *153*
    satellite-based *59, 64, 153, 191*
    seine-netting *136, 137, 196*
    ship-based *64, 166*
    spawning stock *136*
    trawling *136, 137, 196*
sustainability
    use of the sea *184*
sustainable
    development *9,* **10,** *39, 46*
    yield *165, 186*
systematics *106*

T

tagging, radio *90*
taxonomy *15, 17, 112,* **145**
teflubenzuron *95*
temperature *20, 23, 38, 134, 152, 182, 195, 199*
terminology of monitoring *176*
tidal waters **180,** *197*
    prortection of *184*
tides *42*
tiered testing 194
tin, tributyl *47*
tourism *67, 184, 187*
toxicity *147*
toxins *112,* **159**
tracking *81, 90, 187*
tradition, oral *149*
traditional activities *165*
Traditional Ecological Knowledge *149, 155*

Tragedy of the Commons **186**
training *196*
transects *15, 88*
transport *58, 64*
tree rings *27*
trends in quality, tracking of *177, 181, 182, 187*
trophic state *26, 67, 180*
tsunami *56*
turbulence *42*

U

ultra-violet radiation *1*, **42**
uncertainty *1, 155, 191*
    management of *192*
    reducing *189, 192*
United States of America *75, 132, 159*
    Weather Service *191*
Universities
    Harvard *186*
    Stanford *186*
uranium, depleted *53*

V

Valentia Observatory *39*
variability *28*
    man-induced *188*
    natural *2, 19, 50, 188,* **193**
    seasonal **119,** *189*
Viking ships *110*
vineyards, records of *11*
viruses *187*
volcanoes *56*
volunteers, use of *12, 195*

W

Wales *46*
    Anglesey *41*
walking, hill- *61*
waste *49*
    management policy *187*
    sites, hazardous, clean-up *60*
    treatment of *11*
wastewater, treatment of *47, 160, 181, 187*
water/s *54*

bathing *7*
bottled *54*
column *108, 141*
-contact activities *187*
drinking *7*
management *194*
quality *8,* **37,** *75, 135, 177, 181, 187, 195, 196*
    classification *182*
    river *182*
resources *48, 59*
river *181*
    temperatures *182*
sources, protection of *59*
status of, under the Water Framework Directive *72*
tidal *180*
    eutrophication in *37*
    protection of *184*
transitional *176, 177, 178*
transparency *67, 77*
turbidity *42*
Water Education for Teachers, project (USA) *195*
Water Framework Directive **8,** *22, 48, 59, 72, 74, 77, 175,* **176,** *177, 178, 188*
waves *117, 197*
weather and weather patterns *9, 12, 38, 43, 62, 162, 186, 191*
Weather Service, USA *191*
whales *45, 186, 188, 199*
    colouration *199*
    DNA in *186*
    management programmes *186*
    migration of *162*
    North Atlantic Sightings Survey *166*
    Small Cetacean Abundance in the North Sea (survey) *166*
    tail flukes *199*
    tracking *199*
wind *152*
World Health Organisation *7*
World Wars *53*

Y

young-of-the-year index and surveys (fish) *133, 136, 200*

Z

zooplankton *42*

b) People

(other than authors contributing to or cited by reference in this book)

Balech, Enrique *112*
Cabot, David *81*
Carson, Rachel *47*
Darwin, Charles *18, 41*
Faraday, Michael *40*
Farman, Joe *52*
Frank, Ken *151*
Fuchs, Vivian *42*
George, Glen *41*
Hardy, Alister *7, 41*
Jefferies, Don *90*
Jones, Eifion *41*
Laws and Gilbert *41*
Lomborg, Bjørn *20*

Lovelock, James *53*
Lund, John *41*
Marsham, Robert *12*
Mitchell Jones, Tony *90*
Murphy, Matt *36, 112, 130*
Palumbi, Stephen *186*
Roman, Joe *186*
Ruttledge, Robert *81*
Thatcher, Margaret *43*
Virgil *6, 30*
von Humboldt, Alexander *40*
White, Gilbert *6*
Worcester, Robert *51*

---

c) Species

A

Alca torda 84
Alexandrium fundyense 157, 158
algae, blue green (cyanobacteria) 69
alligator 47, 51
Alosa sapidissima 137
Anabaena sp 66
Anas rubripes 158
Anguilla anguilla 89
aphids 7
Asterionella 41

B

Balaena mysticetus 162
Balaenoptera
    acutorostrata 162
    bonaerensis 162
    borealis 162
    edeni 162
    musculus 162
    physalus 162
barnacle *117*, **118**, *120, 123, 130*
bass 76
    Atlantic striped, see *Morone saxatilis*
    striped, see *Roccus saxatilis*
birds (general) 10, 47, 184, 188
bivalve molluscs 153
bladder wrack, see *Fucus vesiculosus*
Branta bernicla hrota 79, 80, **81**
bull huss 76
butterfish, see *Pholis gunnellus*

C

Cepphus grylle 84
Cetaceans **161**
char, see *Salvelinus alpinus*
Charophytes 68

chough, see *Pyrrhocorax pyrrhocorax*
Chrysochromulina 107
Cladophora rupestris 119
clams, soft-shelled, see *Mya arenaria*
cod 185
    Western Atlantic stocks 193
coot 80
copepods *93, 95*
cordgrass, see *Spartina alterniflora*
cormorants, see *Phalacrocorax carbo*
corncrake, see *Crex crex*
Crex crex 80
crustaceans, general **149**
cuckoo *12*
cyanobacteria 69

D

dab 76
Delphinus delphis 161
Dinophysis 107, 143, 147
    acuta 143, 144, 145, 146, 147
    norvegica 147
divers 80
dog whelk **47**, **118**, *120, 121, 124*
dolphins *161, 199*
    bottlenose *170*
    short-beaked common, see *Delphinus delphis*
    white-beaked, see *Lagenorhynchus albirostris*
duck 80
    black, see *Anas rubripes*
    eider, see *Somateria mollisima*

E

eel, see *Anguilla anguilla*
Enteromorpha 81, 119
Eschrichtius robustus 162
Eubalaena
    australis 162

glacialis 161, 162
japonica 162

F

Falco peregrinus 12, 80
Festuca 81
flatfish *89*
flounder *76, 191*
Fratercula arctica 84
fucoids 125
Fucus
    spiralis 120, 127
    vesiculosus 120, 127, 128
fulmar, see *Fulmarus glacialis*
Fulmarus glacialis 84

G

gadoid *89*
gannet, see *Morus bassanus*
Gibbula umbilicalis 128
Gonyaulax tamarense 158
goose *80*
    pale-bellied Brent, see *Branta bernicla hrota*
grebes *80*
guillemot, *86*
    Common, see *Uria aalge*
    Black, see *Cepphus grylle*
gulls *80, 147*
    Great Black-backed, see *Larus marinus*
    Herring, see *Larus argentatus*
    Lesser Black-backed, see *Larus fuscus*
greenfly, see aphids
Gymnodinium
    catenatum 110
    mikimotoi 141
Gyrodinium aureolum 141

H

haddock, see *Melanogrammus aeglefinus*
heron *80*
Homarus americanus 153, 154
house martins *12*
Hyperoodon ampullatus 161

K

Karenia
    brevis 111
    mikimotoi 141
killer whales, see *Orcinus orca*
kittiwake, see *Rissa tridactyla*

L

Lagenorhynchus albirostris 161
Larus
    *argentatus 84, **85***
    fuscus 84
    marinus 84
limpets **118,** *120, 121, 124, 125*
Littorina
    neritoides 128
    obtusata 120, 130
Lobster, American, see *Homarus americanus*

Lomentaria articulata 119

M

Mammals, marine *154*
Manx Shearwater, see *Puffinus puffinus*
Megaptera novaeangliae 162
Melanogrammus aeglefinus 153
molluscs, general **149**
monkfish *76*
*Morone saxatilis **132***
Morus bassanus 84, **85**
mullet *76*
mussel *117,* **118,** *120, 123, 143*
    blue, see *Mytilus edulis*
Mya arenaria 158
Mytilus edulis 158

N

Nucella lapillus 124

O

Oncorhynchus mykiss 17, 92, 99
Orcinus orca 161
Osmundea pinnatifida 119
oyster *47, 143, 187*

P

Patella 125
pepper dulse, see *Osmundea pinnatifida*
peregrine falcon, see *Falco peregrinus*
periwinkles *118*
    flat *120, 125*
Phalacrocorax
    aristotelis 84, **85**
    carbo 80, 84
Phocoena phocoena 161
Pholis gunnellus 89
Physeter macrocephalus 162
pipefish *89*
Placopecten magellanicus 153, 162
plaice *76*
porpoises *161*
    harbour, see *Phocoena phocoena*
Puccinellia 81
Puffin, see *Fratercula arctica*
Puffinus puffinus 84
Pyrrhocorax pyrrhocorax 80

R

ray
    blonde *76*
    homelyn *76*
    painted *76*
    sting *76*
    thornback *75, 76*
    undulate *76*
Razorbill, see *Alca torda*
Rissa tridactyla 84
roach, see *Rutilus rutilus*
Roccus saxatilis 199, 200
rudd, see *Scardinius erythrophthalmus*
Rutilus rutilus 70, 77

S

Salmon,
    Atlantic, see *Salmo salar*
    Pacific Coast stocks *197*
Salmo
    gairdneri 17
    salar 92
Salvelinus alpinus 69, 77
scallop,
    bay *192*
    sea, see *Placopecten magellanicus*
Scardinius erythrophthalmus 70, 77
scorpion fish *89*
sea lice,
    Caligus elongates 92
    Lepeophtheirus salmonis 92, 99, 100, 101, 104
sea urchin, see *Strongylocentrotus droebrachiensis*
seal, *15*, *189*
    grey *152*
sea-lion *189*
seaweeds, brown *88*, *120*, *122*
shad, American, see *Alosa sapidissima*
shag, see *Phalacrocorax aristotelis*
shark
    blue *75*, *76*, *78*
    porbeagle *76*
    thresher *76*
shellfish, general *143*
skate
    common *76*
    long-nosed *76*
    white *76*
smooth-hound *76*
Somateria mollisima 158
Spartina alterniflora 197
Spinachia spinachia 89
Sterna
    dougallii 82, 83
    hirundo 83
    paradisaea 83
stickleback, sea, see *Spinachia spinachia*
Strongylocentrotus droebrachiensis 154
sun-fish *76*
swallows *12*, *45*
swan *80*

T

tern *80*
    Arctic, see *Sterna paradisaea*
    common, see *Sterna hirundo*
    Roseate, see *Sterna dougallii*
tope *75*, *76*
topshells *118*
trout, *77*
    brown *66*
    rainbow, see *Oncorhynchus mykiss*
    sea **72**, *73*
tuna, Atlantic blue-fin *188*
turtles, sea *189*

U

Ulva 81
    lactuca 119
    aalge 84, 86

W

Waders *80*
whales *7*, *161*
    Antarctic minke whale, see *Balaenoptera bonaerensis*
    blue whale, see *Balaenoptera musculus*
    bowhead (or Greenland right whale) *Balaena mysticetus*
    Bryde's whale, see *Balaenoptera edeni*
    common minke whale, see *Balaenoptera acutorostrata*
    fin whale, see *Balaenoptera physalus*
    gray whale, see *Eschrichtius robustus*
    humpback whale, see *Megaptera novaeangliae*
    North Atlantic right whale, see *Eubalaena glacialis*
    north Pacific right whale, see *Eubalaena japonica*
    northern bottle-nosed whale, see *Hyperoodon ampullatus*
    Sei whale, see *Balaenoptera borealis*
    southern right whale, see *Eubalaena australis*
    sperm whale, see *Physeter macrocephalus*
Wrasse *89*

Z

Zostera 81, 82